REFIGHTING THE

PACIFIC WAR

REFIGHTING THE PACIFIC WAR

An Alternative History of World War II

Edited by Jim Bresnahan

NAVAL INSTITUTE PRESS
Annapolis, Maryland

Naval Institute Press
291 Wood Road
Annapolis, MD 21402

First Naval Institute Press paperback edition published in 2023.
ISBN: 978-1-55750-166-0 (paperback)
ISBN: 978-1-61251-068-2 (eBook)

The Library of Congress has cataloged the hardcover edition as follows:
Refighting the Pacific War : an alternative history of World War II / edited by Jim Bresnahan.
p. cm.
Includes bibliographical references and index.
ISBN 978-1-59114-079-5 (hardcover : alk. paper) — ISBN 978-1-61251-068-2 (ebook) 1. World War, 1939–1945—Pacific Area. 2. World War, 1939–1945—Campaigns—Pacific Area. I. Bresnahan, James C., date
D767.R398 2011
940.54'26—dc23

2011018565

∞ Print editions meet the requirements of ANSI/NISO z39.48-1992 (Permanence of Paper).
Printed in the United States of America.

31 30 29 28 27 26 25 24 23 9 8 7 6 5 4 3 2 1
First printing

All photos are from the U.S. Naval Institute Photo Archive.

Contents

Preface and Acknowledgments

WORLD WAR II HISTORY has always fascinated me, especially the Pacific naval war. However, I have often wondered whether the war could have unfolded differently. Was it inevitable? Could Japan have ever won? What if the Japanese had triumphed at Midway or Guadalcanal? Was the atomic bomb necessary? These are just a few of the questions addressed in *Refighting the Pacific Naval War*, as historians, authors, and veterans refight the war while exposing us to a broad range of views.

Refighting the Pacific Naval War focuses largely on naval actions. While I included island battles like Tarawa, Saipan, Iwo Jima, and Okinawa, it is beyond the scope of this work to cover in detail both naval and ground warfare, let alone events involving the British, Dutch, and Chinese. I apologize for not being able to devote more space to all aspects of the Pacific War.

Without a doubt, the key to counterfactual history is maintaining a short-term focus. It's one thing to speculate how the Japanese could have won at Midway; it's quite another to turn such a victory into a successful prosecution of the entire war. If used properly, the counterfactual approach can educate us about what happened and what *could have* happened.

I think most of us want to know if history could have turned out differently. While this book may start more arguments than it settles, if the panelists can get us thinking about how history might have gone down a different road, then *Refighting the Pacific Naval War* has accomplished its goal.

As editor, I recognized early on that this project could not succeed without the participation of some of the top Pacific War experts, the historians and authors that keep alive the memory of the conflict. I also need to thank Vice Admiral Yoji Koda, Japan Maritime Self-Defense Force (Ret.), for providing invaluable commentary in the book's introduction. I also appreciate the handful of veterans who graciously made themselves available for interviews.

Let me add that some of the commentary came from written submissions and some through the transcription of interviews. I applaud all of the panelists for their willingness to speculate on how the Pacific War might have transpired.

Thanks to Ronald W. Russell and Jehanne Moharram for proofreading the manuscript, providing corrections and editing suggestions. A big thank you also goes out to the staff of the Naval Institute Press, including Tom Cutler and Emily Bakely, for their editing and organizational skills and for their decision to explore the realm of counterfactual history.

I also want to thank my wife Mercy and my family for their support, and I praise God who sustains me.

I dedicate this book to World War II veterans and to veterans everywhere, past, present, and future!

Jim Bresnahan
Lexington, Virginia
March 2011

Abbreviations

CAG	commander, air group
CAP	combat air patrol
CinC	commander in chief
CINCPAC	commander in chief, Pacific Fleet
CINCUS	commander in chief, U.S. Fleet
COMINCH	commander in chief, U.S. Navy
COMSOPAC	commander, South Pacific Area
F4F	Grumman Wildcat fighter
FDR	Franklin Delano Roosevelt, U.S. president
FEAF	Far East Air Force
GHQ	general headquarters
GOJ	government of Japan
HIJMS	His Imperial Japanese Majesty's Ship
HQ	headquarters
IJA	Imperial Japanese Army
IJAAF	Imperial Japanese Army Air Force
IJN	Imperial Japanese Navy
JCS	Joint Chiefs of Staff
JMSDF	Japan Maritime Self-Defense Force
LCVP	landing craft, vehicle and personnel
LOCs	lines of communication
LVT	landing vehicle, tracked
NGS	Navy General Staff (Japan)

OTC	officer in tactical command
PACFLT	Pacific Fleet (U.S. Navy)
PBY	patrol bomber (known as "Catalina"; manufactured by Consolidated Aircraft)
POW(s)	prisoner(s) of war
PT boat	patrol torpedo boat
RAAF	Royal Australian Air Force
SBD	ship-borne dive-bomber (Douglas Dauntless)
SCAP	Supreme Commander of the Allied Powers
SEALs	sea, air, and land maritime Special Forces
SNLF	Special Naval Landing Forces (Japanese)
SWPA	Southwest Pacific Area
TBD	torpedo bomber Devastator (manufactured by Douglas)
TF	Task Force
TG	Task Group
USG	U.S. government
USN	U.S. Navy
USSBS	United States Strategic Bombing Survey
VB	bombing squadron
VF	fighting squadron
VMSB	Marine dive-bomber squadron
VS	scouting squadron
VT	torpedo squadron
YE-ZB	tracking/navigation system guiding planes back to their ships/fleet

Introduction

JAPAN FOUGHT A WAR AGAINST the United States and other allied nations in the Pacific, and the result was devastating and humiliating for Japan. Several key reasons for the Japanese decision to conduct that war, in unfavorable situations, have been repeatedly told and published. However, I think it is still worth re-examining several key points of the war.

This ambitious and excellent book challenges us to re-think these subjects from an American point of view. I think it is also important to tell the story from the Japanese side. As an ex–commander in chief (CinC) of Japan's Self-Defense Fleet, which was functionally the same billet as Admiral Isoroku Yamamoto, I think I am in a position to present some views and tell stories of the Japanese side.

In 1940, there was a governmental research organization called the "Soryoku-Sen Kenkyu-Syo" (Total War [Soryoku-Sen] Research Institute [Kenkyu-Syo]), located in Tokyo. This institution was tasked with conducting precise research and analysis on what a "total war" would involve.

The government of Japan (GOJ) estimated that the prospect of a war with the United States was an immediate concern and might be unavoidable. So GOJ tried to prepare Japan for a total war. Based on this outlook, the Imperial Cabinet of Japan, led by the prime minister, Prince Fumimaro Konoe, established the Soryoku-Sen Kenkyu-Syo, and in spring and summer of 1941, wargaming on a hypothetical war with the United States was conducted.

The results of the games were clear and striking. Japan would successfully gain victory and maintain its war momentum in the early phase of the war.

However, as the war continued, it would inevitably become prolonged, and Japan, which lacked almost all the critical natural resources, would not be able to sustain combat operations. As a result, Japan's war effort would start stalling, and eventually Japanese forces would be beaten on every front by U.S. forces. Additionally, in the last phase of the war, the Soviet Union would most likely join the United States and fight against Japan. Thus, the defeat of Japan was clear and inevitable.

Interestingly enough, this analysis, which was developed three months before the outbreak of the Pacific War, precisely forecast the real progress of World War II in the Pacific, including the strategic bombing of Japan. The only events it did not foresee were the surprise attack on Pearl Harbor and the use of nuclear weapons. The Soryoku-Sen Kenkyu-Syo's conclusions were presented to the prime minister and other Japanese leaders on August 27 and 28, 1941. The degree of influence this report had on the leaders of Japan at that time is unknown.

During the years leading up to 1941, the "silent majority" of ordinary Japanese citizens knew that war with the United States could become extremely difficult, and the odds were against Japan achieving a total victory. GOJ and its military also hoped to avoid war with the United States. Even in the 1930s, when Japan was bogged down with prolonged military operations in China, Japan thought that it was still possible to avoid the war.

For Japan, the disagreement with the United States and the root cause of the potential conflict was not a direct competition over national sovereignty, or a territorial dispute, or a clash of fundamental national interests of the two nations; rather, it was an indirect confrontation over a third country—China. In the end, for GOJ, a war with the United States could have been avoided if Japan had successfully contained the conflict within China.

However, the reaction of the United States was very different. Initially, the U.S. government (USG) was patient, and its reactions to Japan were well controlled. However, as violence in China escalated and Japan expanded its operations toward the south, the American people's dislike of Japan grew steeply, in proportion to the worsening situation. USG started taking a hard-line position toward Japan, freezing Japanese assets in the United States, issuing sanctions on industrial materials, and finally imposing an oil embargo in 1941. These were reactions of the United States to Japanese activities in China, which could no longer be tolerated.

So if Japan kept on expanding toward the south of East Asia for natural resources, the worst scenario for Japan would occur—that is, war with the United States. However, Japan's growing concern in the late 1930s and into 1940 was the prolonged military conflict in China. The conflict was practically becoming a total war, which required the mobilization of all possible resources of the country. At this point, GOJ had three difficult issues to handle simultaneously: one was to resolve the bogged-down military conflict in China; the second was to ease growing tensions with the United States caused by the conflict in China; and the last was to prepare Japan for an unthinkable a war with the United States.

It was truly a complicated time in history, and within this excellent book, several questions are raised on Japanese strategy in this period. For example, knowing the huge difference in national power between Japan and the United States, what was the strategy and policy of GOJ for going to war with the United States? Or, how did Japan prepare itself for such a war? The following thoughts on Japan's position may help in answering some of these questions.

Japan and its military started to look at the United States as a hypothetical enemy after the end of the Russo-Japanese War in 1905. Various military strategies and plans—which were equivalent to the U.S. Orange and Rainbow War Plans—were developed and matured by the late 1930s, and they contained several key tenets and assumptions, as follows.

The first priority for Japan was to make the war as short and limited as possible. This reflected the lessons of the Russo-Japanese War and the Great War (World War I), as well as the fact that Japan was still a relatively weak industrial nation, with poor national resources. Secondly, key objectives of the war should be established in the shortest possible time by winning several decisive battles. Finally, as for termination of the war, an advantageous reconciliation, rather than a perfect victory, should be pursued and established at the earliest opportunity.

In order to realize this strategy, Japan's army and navy would need to gain initial victories and maintain their advantages until the United States agreed to come to the negotiating table to end the conflict. The degree of victory in the initial combat actions would need to be decisive and prominent enough to discourage Americans of their will to continue the war—a 1940s version of "shock and awe," so to speak. This was seen as the only way to bring the war to a desirable end for Japan.

In order for the Imperial Japanese Navy (IJN) to fully meet the difficult challenges of this strategy, the Navy General Staff (NGS) had developed a plan to engage the U.S. Navy's Pacific Fleet (PACFLT) from the early 1910s through the 1930s. Its name was the "Gradual Attrition Strategy."

The outline of the strategy was, first, to gain information on the operations and deployments of PACFLT, through wide-area surveillance conducted by IJN aircraft and submarines. Based on their collected information, the IJN submarine force would conduct repeated attacks on PACFLT. Then IJN land-based naval aviation, deployed on various Pacific Ocean islands, would make airstrikes on PACFLT. NGS calculated that these initial engagements would cause a certain amount of damage and casualties within PACFLT. Thereafter, if PACFLT continued to deploy to the west, IJN cruiser and destroyer forces would conduct torpedo attacks, preferably at night. Through these surface torpedo actions, the strength of PACFLT's battleships should be reduced to the same level of the IJN Combined Fleet. At that point, an all-out engagement between battleships would commence to finish the combat actions of the war. IJN had a strong confidence in its tactics, and in the skill levels of its sailors and aviators and their ability to carry out this plan. The IJN Combined Fleet would gain an overwhelming victory through the full annihilation of PACFLT, which Japan hoped would then make the war a limited one and set the stage for truce negotiations with the United States.

However, the strategy of the United States was very different in reality, and the actions taken by the United States did not match the assumptions and expectations of IJN. It was natural that the United States and its PACFLT would develop plans that countered and defeated Japan's strategy. The United States, as a huge industrial nation, wanted to pursue a total war by taking full advantage of its more robust matériel and resource capabilities. In addition, U.S. policy on war termination was not reconciliation, but rather the unconditional surrender of Japan—which the United States believed was possible because the Japanese military could surely be destroyed over time by overwhelming U.S. military power.

As it turned out, the reality of the Pacific War was completely different from Japan's estimates. The most fatal miscalculation by Japan was the fact that the war became a prolonged, total war of attrition, which was what Japan least wanted. To better understand how GOJ could develop such tragically mistaken strategies and war plans, it is necessary to understand several key historic incidents, which influenced the Japanese leaders and decision makers of that era.

First is the lesson of the Russo-Japanese War. There was one thing that should not be overlooked in this war, namely, the "Triple Intervention" by Russia, France, and Germany, which occurred in 1895, immediately after the end of the Sino-Japanese War. The response of GOJ to this intervention was extremely thoughtful: it avoided a military response, and instead integrated its national power to prepare for a future war against Russia. Japan calculated every element of international power and national capabilities, and then waited ten years until it became ready to fight Russia. This provides a striking contrast to GOJ actions in the Pacific War.

Additionally, in the Russo-Japanese War, IJN's NGS and Combined Fleet Headquarters developed a strategy and tactics to gain a "grand slam" victory over the giant Russian naval fleet. In this strategy, IJN, whose main fighting force was a single combined fleet, tried to destroy two large Russian fleets, one after the other. The simple principle behind this was to destroy Russia's Pacific fleet before the arrival of its Baltic fleet as reinforcements. The concept of destroying enemy forces in sequence was a sound strategy for a weaker force, and became the basis of the Gradual Attrition Strategy.

Secondly, there is the influence of World War I. Japan did not participate in the major World War I land battles in Europe. Japan only joined the war by conducting smaller operations and battles in the Far East and the Mediterranean Sea. As a negative consequence, Japan failed to grasp the cold reality and human and resource costs of total and prolonged war. With regard to military operations, IJN sent many war observers to Europe. Combat lessons provided by them influenced the development and planning of the Gradual Attrition Strategy. Three major lessons included: (1) the importance of big-gun engagements, (2) the growing potential of naval air power, and (3) the submarine as an important and lethal force in fleet operations. Unfortunately, IJN missed a key lesson about the importance of protecting merchant shipping, over which Great Britain, as an island empire (similar to Japan), struggled so terribly during World War I.

Of note, Japan gained some dividends from World War I. One was to gain possession of former German territories in the Far East and the Pacific, such as the Marianas, the Carolines, and the Marshalls, which became Japanese mandates. Another, lesser-known benefit was the capture of several modern German submarines and their transfer to Japan. Based on this, IJN was able to analyze and incorporate advanced and complex German submarine technologies into its own force. These two dividends also contributed to the planning and development of the Gradual Attrition Strategy. IJN submarines were expected to play an important role in ambush operations. The widely dispersed, formerly German-occupied Pacific islands, though their fortification was prohibited by the peace treaty, would be potential sites for air bases to be used by land-based naval strike and fighter aircraft.

A third international development that influenced Japan was the Washington Naval Treaty of 1922. After the end of World War I, IJN undertook an ambitious force-building plan. However, at the same time, the GOJ budget situation was very critical. For example, the naval construction budget in 1920 was estimated at 32 percent of the nation's annual budget, which was extremely large for peacetime. In contrast, in 1903, immediately before the Russo-Japanese War, naval construction was only 15 percent of the GOJ's budget. With this situation in mind, GOJ and IJN participated in the Washington Naval Conference of 1921–22. The negotiations produced the concept of IJN maintaining a 60 percent ratio of capital ship tonnage, compared to the U.S. Navy and Britain's Royal Navy. There was strong opposition to this within IJN; however, the extraordinary leadership and determination of Japan's navy minister, Admiral Tomosaburo Kato, who also was a chief of Japan's delegation to the conference, enthusiastically integrated IJN opinions, and the treaty was successfully concluded. Also, in hindsight, it is now known that there were crafty U.S. strategic maneuvers and code-breaking occurring behind the scenes during the negotiations, but these should not take anything away from the achievement of Admiral Kato.

A 60 percent ratio of capital ship tonnage was a serious limitation, but considered to be manageable by the NGS. International cooperation and Japanese domestic financial factors had clearly overridden the ambitions of IJN. In other words, what Japan gained from the Washington Naval Treaty was a period of stability among the major naval powers. At the same time, however, it was true that dissatisfaction toward the 60 percent ratio among some IJN officers persisted beneath the surface. Unfortunately for Japan, this un-extinguished ember would blaze up into a bursting flame of political dispute at the conclusion of the London Naval Treaty in 1930. Similarly, GOJ traded one key thing away for this treaty, in that it terminated the Anglo-Japanese Alliance, which had been a core element of its diplomacy and security for twenty years. Japan would have to pay the cost for its broken tie with the United Kingdom in later years.

A fourth influence was the London Naval Treaty of 1930. This treaty created irreparable confusion and trauma, both in IJN and Japanese politics. The main subject of the conference, ratios of cruiser forces and other auxiliary assets among the three major navies—IJN, the U.S. Navy (USN), and Britain's Royal Navy—was an important issue in Japan. Unfortunately, there was a serious disagreement between GOJ and NGS. GOJ put higher priority on international cooperation, and worked to determine if the ratio could be acceptable to IJN. The eventual decision of GOJ and the Navy Ministry was to accept the treaty negotiation result because the agreed ratio of 0.6957 was little different from GOJ's original proposal of 0.7. However, the NGS obstinately opposed GOJ. From the position of the NGS, the limitations of the London Treaty would narrow the capabilities of IJN's Combined Fleet, especially the capabilities of its torpedo forces—that is, heavy cruisers and submarines—and as a consequence, it would lose the war against USN's PACFLT. A serious fissure between the Navy Ministry and NGS generated deep psychological separation within IJN for the first time.

An important issue related to this was the *tosuiken kanpan mondai*—the possible violation of the prerogative of supreme command of the emperor, which was raised during the ratification process of the London Treaty. During ratification, the opposition political party attempted to use this navy issue as a political killer punch aimed at the ruling party. The opposition party asserted that approval of the London Treaty, without the confirmation of the NGS, was a violation of the constitution of Japan. Article 11 of Japan's Meiji Constitution defined the prerogative of supreme command of the emperor, which was called *tosuiken*. The opposition party claimed that authority to decide the size and strength of the military services fell under *tosuiken*. Therefore, the conclusion of the London Treaty without the agreement of the NGS was unconstitutional. This political ploy caused a significant impact, both inside and outside of the Diet (Japan's national parliament). For the first time, the extremely sensitive issue of the constitutional authority of the emperor was willfully used, in an unhealthy way—that is, to simply attack the ruling party government. However, in the end, the majority of the Japanese people supported the London Treaty, the willful attempt of the opposition party collapsed, and the treaty was ratified by an overwhelming majority at the Privy Council in December 1930.

This type of political chaos and its resultant psychological divisions within IJN left deep scars that did not heal. This became an indirect cause for GOJ's loss of control over the military in the 1930s. The Japanese military, especially the Imperial Japanese Army (IJA), learned one evil lesson through this political confusion. The political tactic that IJA learned was that any military issue, especially operational matters and force planning, could fall under *tosuiken*; therefore, the other GOJ ministries and agencies could not intervene. This was not a correct tenet under the constitution; however, the political community in Japan failed to effectively oppose it. After that, IJA realized all of its policies and initiatives under the authority of *tosuiken*, and GOJ, in effect, lost control over IJA. Another infamous IJA trump card was to effectively veto cabinet decision making by not sending the army minister to attend its session. The main result of the above factors was the runaway, out-of-control IJA of the 1930s.

Here I would like to switch from the history of strategy and policy to naval warfare.

Since naval aviation was a main factor in the Pacific War, it is necessary to examine IJN naval aviation more closely. Japan started its military use of air space in 1909, six years after the first flight of the Wright brothers. After the first flight of fixed-wing airplanes in 1912, IJN aircraft were deployed during the operation to seize German-occupied Qingdao (known in the West as Tsingtao) during World War I.

However, the naval aviation and industrial capability of Japan remained in the "infancy" stage during the following years, and this led IJN to invite a British team for assistance. Great Britain sent its mission, led by Sir William Francis Forbes-Sempill (an ex–Royal Air Force officer), to Japan. The team stayed in Japan in 1921 and 1922. They taught IJN all the elements of naval aviation, such as the latest tactics and operations including aerial torpedo attack methods. As a result, the visit of the Sempill mission became a baseline and building block for improving IJN naval aviation.

IJN built its first aircraft carrier in 1922. HIJMS *Hosho* was the first ship in the world constructed as an aircraft carrier from design to completion. As for carrier landings, the first carrier landing on *Hosho* was successfully completed by an IJN pilot in March 1923.

As agreed in the Washington Treaty, HIJMS *Akagi* and *Kaga* were converted from a battle cruiser and a battleship, respectively. Originally, two fast (32 knots) battle cruisers, *Amagi* and *Akagi*, were to be converted, but a giant earthquake in 1923 badly damaged *Amagi* at Yokosuka Naval Shipyard. So *Kaga* (26 knots) replaced *Amagi* and completed her conversion in 1928. With the completion of the smaller HIJMS *Ryujo* in 1933, IJN made steady progress in all areas of naval aviation. By the mid-1930s IJN had strong confidence in its aviation capabilities. In 1934 and 1935, *Kaga* underwent a major modernization, including the installation of a single through-flight deck and an island on the starboard side. Her maximum speed was increased to 28 knots. *Akagi* followed with similar upgrades in 1938. She reflected all the previous technical and design lessons, except with her island. Due to diverse opinions among aviators, the island of *Akagi* was placed on her port side for comparison.

In 1937 and 1939, two new fleet carriers, HIJMS *Soryu* and *Hiryu*, joined the IJN fleet. They were medium carriers, with identical designs; however, *Hiryu* was completed as a slightly different ship, including a portside island. These two carriers were the first carriers that fulfilled the operational requirements of IJN. Then, two much improved sisters, HIJMS *Shokaku* and *Zuikaku*, went into service in 1941. These two ships were real twin sisters, both with starboard islands. They displaced 32,000 tons each at full load, with 242-meter-long (794-foot-long) flight decks. Their maximum speed was 34 knots, and each of them could carry more than seventy aircraft. These ships were the best aircraft carriers in the world, until the commissioning of USS *Essex*.

As for aircraft manufacturing, after a twenty-year-long dependence on foreign technologies, Japan started a full domestic development and production posture in the latter half of the 1930s. Rear Admiral Isoroku Yamamoto, who was a director of the aviation technology department at Navy Aviation Headquarters, was a strong advocate of this policy. In the latter half of the 1930s, together with new carriers, newly developed carrier aircraft, such as the Type 96 fighter (Claude), the Type 96 dive-bomber (Susie), and the Type 97 attacker (Kate), started entering fleet service. The Type 99 dive-bomber (Val) in 1939 and the Type 0 fighter in 1940 followed their predecessors.

IJN had developed a unique operational concept of land-based aviation. IJN specially designed and built land-based strike and reconnaissance aircraft as major players in fleet operations, to realize its Gradual Attrition Strategy. IJN also directed its keen attention to the rapid improvements of aircraft in the 1930s, including: long-range navigation and communication capabilities, upgraded payload for aerial torpedoes, and an expanded operational radius. A key to the birth of land-based attack aircraft were the widely dispersed Japanese mandate Pacific islands, which were ideal air bases for pursuing the Gradual Attrition Strategy. To fully realize this idea, the Type 96 attack aircraft (Nell) was developed and entered fleet service in 1936. Nell was a major attack aircraft, and deployed to China as a bomber as well. Without Nell, Japanese ground operations in China could have been very difficult. A Type 1 attack aircraft (Betty) joined the fleet in 1941 as follow-on.

Of note, there was one Achilles' heel, which was common to all the aircraft of IJN, and that was poor and insufficient self-protection. In particular, the lack of self-sealing fuel tanks and armor-plate protection for aviators were fatal flaws in a prolonged war with the United States.

IJN emphasized seaplanes too. Shipboard seaplanes were used as main reconnaissance assets. Cruisers were the primary platforms for these seaplanes. The heavy cruisers of IJN normally carried three seaplanes and two catapults each. In addition to them, IJN developed a scouting heavy cruiser for this purpose. HIJMS *Tone* and *Chikuma*, with all 8-inch guns in forward decks, were completed in the late 1930s. A spotting deck for six seaplanes and two catapults occupied the quarterdeck. Each ship carried Type 0 (Jake) three-seat reconnaissance seaplanes. These two ships were assigned to the Kido Butai (mobile task force) in 1941 and 1942, and acted as the "eyes" of the force. IJN also had a dedicated air branch for reconnaissance. Of note, by December 1941 there were 460 pilot officers and 219 reconnaissance officers in IJN.

Last but not least, IJN expected a lot from a large four-engine flying-boat. Wide-area ocean surveillance and long-range reconnaissance were keys to successful operations of the Gradual Attrition Strategy. These seaplanes also could be operated from the Japanese-controlled Pacific islands. The first successful large flying boat, with an approximate 3,500-nautical-mile operational range, the Type 97 (Mavis) entered IJN service in 1937. The flying boat also had a large payload of two tons, or two aerial torpedoes. As a successor, the Type 2 (Emily) was developed, and entered service in 1942.

In this regard, the doctrine and equipment of IJN did not make light of fleet reconnaissance. The Combined Fleet had sufficient reconnaissance seaplanes and flying personnel. The Kido Butai under the command of Vice Admiral Chuichi Nagumo was not an exception. Nagumo, at the Battle of Midway, had sufficient assets for reconnaissance seaplanes on two cruisers. The problem at that time was not the doctrine of IJN but the poor reconnaissance planning specific to Nagumo's headquarters.

IJN successfully developed practical operational concepts for naval aviation—including the use of aircraft carriers—in the latter half of the 1930s, when carrier aviation in the world's other major navies was not yet fully established and while they were still struggling to develop suitable systems and doctrine. Additionally, lessons accumulated from actual combat experience in China helped IJN develop realistic and practical operational aviation doctrine.

By fully taking the new operational doctrine of naval aviation into account, IJN had a determined intent to integrate all available assets into its Gradual Attrition Strategy. The following roles for naval aviation were incorporated into this strategy: (1) long-range surveillance and reconnaissance by large flying-boats to locate and report west-bound U.S. Navy PACFLT; (2) repeated aerial torpedo attacks by land-based aviation, together with, or after that of, submarine torpedo attacks; (3) neutralization of enemy carriers by carrier-bombers and attackers; (4) establishing air superiority over air space of friendly fleet by carrier-fighters; and (5) disabling or sinking several of the enemy's battleships by carrier attackers and bombers.

In the latter half of the 1930s, the operational plan of the Gradual Attrition Strategy had reached its full maturity. Upon the promotion of Admiral Isoroku Yamamoto to CinC, Combined Fleet, in August 1939, he completely changed the operational concepts of this strategy in the following two years. As a leader who actively promoted the modernization of naval aviation, he strongly directed revolutionary new deployment concepts for his carrier force, and successful validation fleet exercises in 1940 gave Yamamoto strong confidence in the validity of his new ideas. These new systems, ideas, and tactics eventually became the basis for Operation Pearl Harbor.

I would like to stop the descriptions at this point, because the operations that followed in late 1941 are well known to all.

Finally, I would like to point out several other issues that influenced the strategic decisions of both Japan and the IJN.

The first issue concerns cooperation between Japan and Hitler's Germany. In the Tripartite Alliance, the possibility of Japan conducting significant military operations along the Far East front was considered to be low. For Japan in 1940 and 1941, maintenance of the Tripartite Alliance was an important issue; however, it had few national interests in Europe or the Soviet Union. For this reason, it was difficult for Japan to develop a convincing rationale to cooperate with Germany. An analogous situation can be seen during World War I, where the Anglo-Japanese Alliance was still an important commitment, but Japan hesitated to participate in the European land war because it had difficulty identifying a vital national reason to send its forces into Europe. Similarly, military cooperation with Germany in the Far East was an acceptable idea, but the fundamental nature of this issue was not where to fight; rather, it was a matter of national interests. Additionally, there was no mutual trust between Japan and Germany. The reality was that GOJ and IJA were fully occupied by the prolonged military operations in China, and were too busy to develop a new strategy for cooperation with Germany.

A similar idea to this was "bypassing the United States." Japan could have pursued a war to repel the European colonial powers from Asia, at the same time giving Japan an opportunity to gain natural resources. In spite of the potential attractiveness of such a war to many Asians, there were several difficult conditions for Japan. In order for Japan to fight against the European colonial powers and liberate the Asian peoples, Japan needed to convince the United States that Japanese operations in China were not also another type of colonialism. One option would have been to fully withdraw from China, but this was an action to which the IJA would never agree. If GOJ had had tight control over IJA, the story could have been different. This was a dark shadow of the runaway army produced by *tosuiken*. So, though this thought was acceptable in theory, it was not realistic or affordable to Japan.

A second issue, which affected Japanese military strategists, is the well-known very poor self-protection of aircraft, and the inadequate damage control of the warships of IJN. The reason for this situation was that, at first, IJN wanted to fight a short war, with only a few major battles, and IJN hoped to gain victory in a decisive battle. If the Combined Fleet could gain victory in a short war, the overall casualties in the limited number of battles would be sustainable and acceptable for IJN. Large casualties could be borne in such a situation. This was the fundamental thought of IJN, and the real reason for its notoriously poor protection systems. There was one more factor, and that was a trade-off between the attack and defense capabilities of each asset. Numerical inferiority mandated by the two naval treaties (1922 and 1930) forced IJN to pursue some trade-offs. In order to make up for quantitative inferiority, IJN built ships and aircraft with qualitative superiority. Also, IJN clearly prioritized attack over defense (and protection). This type of thinking dominated all the operational requirements and development policies of IJN. Ironically, Japan and IJN were subsequently bogged down in a prolonged Pacific War, in which IJN's fundamental and fatal structural problems of poor protection were painfully exposed.

The third and final issue is Japan's views on the key campaigns and battles in the Pacific War. It was clear that battles in the Solomon Islands, lasting over sixteen months from Guadalcanal (August 1942) to Bougainville (November 1943), had gradually undermined the Combined Fleet. IJN experienced its first large losses of carrier aviators in the Battle of the Coral Sea, then the Battle of Midway—however, these losses were sustainable compared to those of two carrier battles in the Solomon Islands. The six-month-long Guadalcanal campaign really exhausted IJN. The casualties among land-based naval aviation in this period were serious and not sustainable. At the same time, carrier aviation lost most of its experienced aviators in the Battles of the Eastern Solomon and Santa Cruz islands. Through operations following Guadalcanal, IJN aviation and destroyer forces of the Combined Fleet finally wore out. In fact, whole carrier air-groups in the Combined Fleet collapsed in the fall of 1943, and since that time the Combined Fleet lost its air-strike capabilities against the U.S. Navy's PACFLT.

IJN was too late to pursue a substantial buildup program for aviators. In fact, IJN did not start a rapid and massive training program until mid-1943. Before that, for example, the number of reserve flight officers who joined the program in 1941 was only forty-three. In 1942, the number was increased to 278. About half of them were not ready in time for the Battle of the Philippine Sea. Finally, 5,199 and then 3,323 college students joined the naval aviation training program in two consecutive courses in mid- and late 1943; 4,726 and 1,954 of them graduated the courses eleven months later as flight officers. They became combat-ready by the time of operations in the Philippines in late 1944 and during the Okinawa operations in April 1945, respectively. Their average number of flight hours at the completion of flight training was about eighty.

In summary, Imperial Japan gradually narrowed its options for international policy action after the Russo-Japanese War. Japan gained possession of the Liaodong peninsula and southern Manchuria in 1931, and this triggered the initial gap between Japan and the United States. Even so, Japanese international policies and actions remained relatively stable during the next two decades.

The first major international separation was the collapse of the Anglo-Japanese Alliance. For Japan, the alliance had been a passport for entrance into the European-dominated "international club," so the loss of the alliance produced many negative impacts on Japan. The Japanese people became inclined to follow more nationalistic directions after that and, as a result, this led the nation toward isolation. This flow of events was accelerated by the London Naval Treaty of 1930 and the Manchurian Incident of 1931. After that, Japan went down in a straight line, on a steep downhill slope, toward the Pacific War.

In the Pacific War, Japan fought in the least expected scenario, a prolonged, total war of huge attrition, and it ended up fighting in a way it was totally unprepared to. Japan and IJN ended up being destroyed by PACFLT. Sailors, including land-fighting units and aviators of IJN, did their best to the extreme. However, the major battles in 1944 and 1945, when IJN lost its striking capability, were like a boxing match between a big, muscled boxer, full of energy, and a completely exhausted and emaciated player. Naval operations after the Battle of Leyte Gulf essentially became a mop-up for PACFLT. Thus, IJN's Combined Fleet faded away by August 1945.

Ironically, this was the path that the Soryoku-Sen Kenkyu-Syo had predicted in the summer of 1941.

I hope this provides a point of reference and adds some perspectives to this excellent book.

Vice Admiral Yoji Koda,
Japan Maritime Self-Defense Force (Ret.)

One

SEEDS OF CONFLICT

THE STORM CLOUDS that engulfed America and Japan in 1941 had been gathering for decades.

In the 1930s, Japan was a nation plagued by economic woes, failed coups, and violent assassinations. Lacking important resources like rubber and oil, the Japanese elected to seize what they did not possess.[1]

First to fall to Japanese aggression was southern Manchuria in 1931. When the League of Nations condemned the invasion, Japan simply vacated the League.[2]

Japan then shifted its focus to China, launching an invasion at the Marco Polo Bridge near Peking in the summer of 1937. By the end of the decade, the Japanese had grabbed vast amounts of territory, captured Peking, and raped Nanking, but the Chinese refused to quit.

Japan's Chinese excursion also faced growing opposition from America. To protect American interests, President Franklin Delano Roosevelt transferred the U.S. Pacific Fleet from San Diego to Pearl Harbor in Hawaii.[3] Then in the summer of 1941, after Japan pressured the Vichy French to allow Japanese troops in French Indochina, Roosevelt fired off economic salvos, including an oil embargo and a freezing of Japanese assets.[4]

Limitations

Well before Japan's foray into China, representatives from the United States, Great Britain, and Japan gathered in Washington in November 1921 in an effort to forestall an arms race. The delegates wound up approving a 5-to-5-to-3 ratio for the future construction of warships for America, Britain, and Japan respectively. The Japanese had sought a 10-to-10-to-7 ratio, which they believed would leave the United States without the necessary strength to wage an offensive campaign against Japan.[5]

The Washington Naval Arms Limitation Treaty, signed in Washington on February 6, 1922, allowed America and Britain to retain fifteen capital ships each while the Japanese kept nine. The agreement permitted the Americans and British to build aircraft carriers up to a total standard displacement of 135,000 tons, compared to 81,000 tons for the Japanese.[6]

Q: *What if there had not been a Washington Naval Arms Limitation Treaty and America, Japan, and Britain had not faced shipbuilding restrictions?*

John Burton

It is difficult to imagine World War II in the Pacific without visualizing carrier task forces and their swarms of fighters, dive-bombers, and torpedo planes exchanging major blows between opposing fleets. It should appear more than somewhat ironic that much of what we remember about the Pacific War would likely have never occurred if not for the series of post–World War I international naval treaties intended to limit the construction of capital warships.

Before these arms limitation discussions began with the Washington Naval Conference of 1922, carrier aviation was mostly a concept on paper. Only the British had operational aircraft carriers: HMS *Argus*, converted from an Italian ocean liner; HMS *Furious*, converted from a heavy cruiser hull; and HMS *Eagle*, converted from the hull of an Argentine battleship. The Royal Navy had another carrier under construction: HMS *Hermes*, its first vessel to be designed as such from the keel up. The United States was about to launch its own experimental carrier, USS *Langley*, which was essentially created through the makeshift addition of a flight deck to the smaller, lighter shell of an outmoded collier. Japan, with substantial technical assistance from Great Britain, had taken the keel of an oiler and transformed its hull into the world's first commissioned, purpose-built aircraft carrier, HIJMS *Hosho*.

At the time of the initial treaty discussions, none of the victorious navies had developed a solid strategy for employment of their new aircraft carriers.

The IJN had made good use of its seaplane carrier HIJMS *Wakamiya Maru* during the invasion of the German colony of Kiaochow in 1914, and subsequently developed sound plans for the tactical deployment of seaplane fighters, bombers, and flying boats. The utility of wheeled aircraft, launched and recovered from flight decks, was accepted only as an article of faith in all three navies.

Wartime demands spurred construction of a large number of ships, many of which were incomplete at the end of the Great War. The immediate result of the Washington Naval Treaty was to change the direction for construction of several battle cruisers and one battleship that were still in some shipyards. To avoid the drastic measure of scrapping some of these vessels, which were substantially complete, each nation converted two large capital ships into aircraft carriers. The United States built *Lexington* and *Saratoga*; Great Britain finished *Glorious* and *Courageous*; and Japan completed *Akagi* and *Kaga*. Interestingly, as a direct result of this diversion in the treaty limitations for surface combatants, the first real "fleet" aircraft carriers were born—and a brand-new naval arms race commenced.

Without the Washington Conference, these six ships would never have been completed as aircraft carriers, and the doctrine for naval aviation in the United States and Japan would almost certainly never have evolved. American and Japanese experiments constrained to the use of *Langley* and *Hosho*, respectively, could not have led directly to the development of any sizeable quantity of "modern" fleet carriers. The IJN would probably have gone on to develop a highly sophisticated seaplane organization to support its projected amphibious landing operations, while the U.S. Navy might have been content to rely upon its large flying boats—used mostly for reconnaissance. Perhaps only the British would have developed useful aircraft carriers with offensive striking power, in which case the Pacific War would have been quite different.

When war finally came—and it was a virtual certainty it was coming by the time the next generation of aircraft carriers was launched during the 1930s—it would have been too late to develop aircraft carrier systems and aircraft. Battleship admirals of the USN and IJN would undoubtedly have contested for control of the seas in a gunnery and torpedo duel, some deadly reincarnation of Tsushima or Jutland. For such a scenario, there is a high probability that the U.S. Pacific Fleet would have been drawn into combat in the western

Pacific—much in the manner postulated by the architects of War Plan Orange. Unrestrained by an arms limitation treaty, Japan's disproportionate allocation of its gross national product to military development would have resulted in far better capital ships for the IJN than a parsimonious Congress would have provided to the U.S. Navy. There are many reasons to believe that the United States would have been beaten severely in such a battle, losing the war and all of its Pacific Ocean territorial possessions.

As it actually happened—even with the treaty limitations—the Empire of Japan managed to develop better torpedoes, better heavy cruisers, and better battleships before the war began. In fact, because both the United States and Great Britain were always forced to consider deployment for a "two-ocean war," the treaty allowance of ship quantities and tonnage were much more favorable to Japan than the negotiated numerical ratios suggest at face value.

Jon Parshall

I think an interesting question revolves around whether the ratios on capital ships and cruisers had been set at 10-to-10-to-7, rather than 10-to-10-to-6. Conventional naval thinking of the time stated that a defending force needed to have 70 percent of the strength of the attacker in order to conduct a successful campaign. The Americans and British deliberately pegged the Japanese at 60 percent of their strength so as to deny them this measure of security. Had we been more willing to concede a 10-to-10-to-7 ratio, it's possible that the virulently nationalist fleet faction within the IJN might have had less room to maneuver in decrying the treaty system, and this might have marginally retarded the Japanese move toward all-out militarism.

On the flip side, though, it seems fairly clear that Japan's move to militarism was largely pre-ordained. If so, giving Japan the ability to build up to a nominal 10-to-7 ratio with the USN might have simply made matters worse for us in 1942–43. The Japanese, of course, cheated on the treaty system in any case, and built their navy to a ratio approaching 80 percent of the USN's pre-war strength. With another 10 percent worth of "wiggle room," they might very well have nearly matched our pre-war strength, if their economy could shoulder the burden of the extra building programs. That's a pretty big "if," though: all the evidence I have seen on the Japanese pre-war economy suggests that they had pretty much "maxed out" their ability to spend any more than they did on weaponry.

Anthony P. Tully

The problem was not the Washington Naval Treaty of 1922, but rather its perceived different treatment of Japan. Had Japan's navy—already powerful, well equipped, impressively trained, and proven in war—been treated as an honored equal, things might have gone differently. As it was, the lack of parity strongly conveyed an impression of Japan being treated like a "step-child" and perhaps, in part, due to racist motives. It was hard for the Japanese to ignore some of the rhetoric commonly heard about Asia, and particularly about Chinese laborers in the American West.

The answer offered here is predicated on the postulate, impossible to prove, but reasonable, that the time to avoid the Pacific War was arguably in the 1920s, not the 1930s. Unless grounds of an international acceptance and community strong enough to offset the rise of the militarists had existed, the events of the 1930s and 1940s would have been extremely difficult to prevent. The 1920s offered that moment. Japan had cooperated well with the Allies in the struggle in World War I. The Japanese were already showing increasing interest in world trade, and famous purchases of warships built in British yards are one example of this.

A desire to restrain an arms race, to avoid a repetition of World War I, was certainly understandable, and the Naval Treaty itself probably did not spell offense. The issue with the Japanese, particularly in light of the somewhat unfair treatment of their military gains of Czarist Russia in the Russo-Japanese War, was one of continuing perceived (and real) discrimination and non-parity. It is very arguable that offering parity to a power not determined to be an enemy was far less risky than creating the climate that ultimately empowered the militarists and their rise.

In many ways, though other opportunities would exist, once 1931 arrived, and after it the events that led to the start of the Sino-Japanese War and the ascendancy of the Kwantung Army faction in politics, the Pacific War became increasingly probable, not least because of the steady disintegration of the democratic process in the Tokyo government from assassinations and nationalism, as well as the pressures of the Great Depression. Hence, the failure to treat Japan, an ally of World War I, more equitably in the 1920s is more important and decisive than might be realized. Arguably, it finds a parallel in the present, where there exists a sentiment perhaps too quick to believe the worst of Chinese intentions, and to prefer to use a confrontational outlook.

Two-Ocean Navy

In June 1940, the U.S. Congress approved a massive naval expansion, the Two-Ocean Naval Expansion Act, with $4 billion set aside for new construction, committing America to building eighteen aircraft carriers, seven battleships, twenty-seven cruisers, more than one hundred destroyers, and over forty submarines.[7] Upon completion, the construction program would relegate the Imperial Japanese Navy to second-rate status.[8]

Q: *What if Congress had not approved an expansion of the U.S. Navy?*

Jon Parshall

We would have been in deep weeds. Simple as that. The basis for our naval strategy in the Pacific was an understanding that in the long term our naval power was going to be insuperable. That supremacy really starts appearing in early 1944, that is, four years after the Two-Ocean Naval Expansion Act. If that act had not been signed when it was, our Navy wouldn't have had the tools it needed to prosecute campaigns in the central and west Pacific until probably early 1945, and the war might have been fought until 1946. And if that had happened, the shape of postwar Japan would have probably contained a much larger Soviet component.

Tim Jackson

Had not Congress appropriated the funds for naval expansion in 1938 and 1940—called the various Vinson bills after Congressman Carl Vinson (D-GA), chairman of the House Naval Affairs Committee—the chances of the United States prevailing in a naval struggle in the Pacific against the Japanese Empire were minimal. Additionally, the lack of sufficient naval forces for convoy and merchant ship escort and antisubmarine warfare in the North Atlantic would have left Great Britain essentially helpless against the German Kriegsmarine submarine onslaught. With ever-increasing numbers of highly capable Type VII and Type IX U-boats coming online in 1942, without a robust U.S. naval force already in the water as of December 8, 1942, the chances of the U-boat commerce-raiding campaign having succeeded are high.

The 1938 appropriation, called the "second Vinson naval bill," authorized a 20 percent expansion of the fleet as well as the acquisition of three thousand new aircraft. When the actual naval appropriation followed in 1939, Congress budgeted for 950 aircraft. Following the 1938 bill, in 1939 Congress appropriated funds for the laying down of the first two battleships of the new South Dakota class—USS *South Dakota* (BB-57) and USS *Indiana* (BB-58)—based on the first Vinson naval bill, or Vinson-Trammel Act of 1934, which authorized building up to Washington and London naval treaty authorized strength. In terms of aircraft carriers, which after Pearl Harbor became the primary capital ship in the Pacific theater, the two appropriations of 1938–39 resulted in the construction of USS *Hornet* (CV-9) and the lead ship of the class destined to become the mainline U.S. Navy combatant of the war, the USS *Essex* (CV-9). Had not these appropriations become law, the U.S. Navy would have faced the post–Pearl Harbor struggle with only two modern battleships—USS *North Carolina* (BB-55) and USS *Washington* (BB-56), six frontline carriers, of which two—USS *Wasp* (CV-7) and USS *Ranger* (CV-4)—had serious operational deficiencies compared to Japanese carriers, and far too few aging cruisers and destroyers.

By 1940 and the collapse of France in the face of the German onslaught of May–June, Congress, led by Vinson, stood ready to appropriate the money to construct a Navy capable of simultaneous and robust operations in both the Pacific and Atlantic, hence the Two-Ocean Naval Expansion Act, or third and fourth Vinson naval bills. Early in 1940, Vinson shepherded through Congress an appropriation for three additional carriers as well as numerous smaller warships. That would result in an 11 percent fleet expansion once fielded. On May 2, he introduced a bill calling for an increase in naval aircraft from three thousand airframes to ten thousand, and an increase in the number of pilots from the 2,602 in current billets to 16,000. On June 3, he ramrodded through Congress a further bill that would appropriate funds for an additional carrier and a further 11 percent increase in fleet strength.

By late June, with the surrender of France, the stage was set for the most dramatic of all Vinson navy bills, the Two-Ocean Naval Expansion Act. Admiral Harold Stark, chief of naval operations, in testimony before Congress, argued for a 70 percent increase in fleet size to meet the challenge of a potential two-ocean struggle with the Axis powers. Allied with Senator David Walsh (D-MA) in the Senate, with only two hours of debate Vinson pushed through the appropriation bill or Vinson-Walsh Act that set the U.S. Navy on course for a massive shipbuilding and aircraft-building program that would come to fruition by late 1943 and through 1944. By early September 1940, American shipyards had under construction 201 naval vessels. By January 1941, multiple shipyards had under construction seventeen battleships, twelve carriers, fifty-four cruisers, eighty submarines, and 205 destroyers. Airplane manufacturers were turning out Wildcat, Dauntless, Catalina, Devastator, and other aircraft types by the hundreds.

With the loss of the Pacific Fleet battle line at Pearl Harbor on December 7, 1941, without the new ship construction and the thousands of aircraft authorized and then appropriated between 1938 and 1940, the U.S. Navy would simply not have had the resources to deter the advance of Japan in the Pacific. The Japanese Empire likely would have been able to consolidate its gains and establish a defensive perimeter so robust that, even had naval appropriation on a massive scale followed Pearl Harbor, it likely would have been too little, too late. Nor would the Navy have been able to supply adequate protection to the convoys to Britain, thus making the German strategy of commerce interdiction causing a British collapse a distinct possibility.

Squeezing Japan

When Japan occupied Saigon and Danang in French Indochina in mid-1941, President Roosevelt (FDR) responded by freezing Japanese assets. FDR also embargoed oil shipments, a terrific blow since Japan imported nearly 90 percent of its oil, a large portion of it from America.[9]

Q: *What if the United States had not implemented economic sanctions against Japan?*

Bill Bartsch

On July 26, 1941, the U.S. government froze all Japanese assets in the United States, in effect blocking all trade with Japan, including the critical petroleum exports on which the Japanese economy—and its war machine—was so dependent. Earlier, in October 1940, President Roosevelt had introduced a ban on export of scrap iron and steel, aimed at hindering Japan's war-making capabilities.

The decision to freeze Japanese assets was not based on unanimous views of Roosevelt's advisors, and he himself was not fully committed to the drastic course of action (in effect introduced by high-level bureaucrats, as shown in

Edward S. Miller's recently published *Bankrupting the Enemy*). The Navy in particular was strongly opposed, anticipating that the Japanese would respond by seizing the oil fields of the Dutch East Indies (and Malaya) to meet their needs.

However, the Japanese had already been planning to occupy the Indies at least as far back as July 1940, when the emperor endorsed a new government policy to set up a "Greater East Asia Co-Prosperity Sphere" that would incorporate, among others, Malaya, the Dutch East Indies, and Indochina, mainly resources-rich colonies of Great Britain, the Netherlands, and France that looked ripe for the picking following the victories of Hitler's troops in western Europe. With the approval of the "southern advance" to implement the policy, a first step was taken in occupying southern Indochina in July 1941. This aggressive act triggered the imposition of an "oil embargo" by the United States, which was followed by oil embargoes by Great Britain and the Netherlands.

While the Japanese recognized that the occupation of southern Indochina ran the risk of war with the United States and Great Britain, the opportunity to become self-sufficient in natural resources overrode such concerns. The oil embargoes only speeded up Japan's decision to mount its southern operations before its oil supplies ran out and foreclosed its war capabilities. By early September 1941, the likelihood of war with the United States, Great Britain, and the Netherlands was now built into Japanese planning, with the decision taken to go to war unless progress were made in negotiations with the United States by early October.

Thus, in effect, the U.S. embargoes were irrelevant to Japanese war planning, except in affecting the *timetable* for them. Indeed, even the War Department's expectation that its late-1941 massive buildup of air power in the Philippines, with B-17s believed capable of striking the Japanese mainland, would serve as a more threatening deterrent to Japanese expansionist plans than oil embargoes made little difference in Japanese planning once the decision for war had been taken. Short of diplomatic negotiations succeeding in meeting the U.S. demand for a Japanese withdrawal from Indochina and China, war was inevitable.

Stephen D. Regan

Franklin Roosevelt did not want war between the United States and Japan if he could avoid it; however, the intransigent and brutal actions of Japan between

1905 and 1941 reached a level to which the president had to react in some significantly diplomatic manner short of war. The chronic coups and attempted coups, the increased dictatorial actions of the army, and the obvious aggression against the United States and Great Britain demanded drastic actions by the West. The blatant sinking of the *Panay* and earlier bombings of British boats, plus the pretentious invasion of Manchuria, paled in comparison to the Rape of Nanking, where over 250,000 civilians were coldly slaughtered. No president or leader of an industrialized nation could make any claim to moral leadership without reacting harshly to Japan. FDR had no choice but to reluctantly impose economic sanctions, including the oil embargo.

In spite of that, Roosevelt was looking at Germany as his primary problem, not Japan. When the Japanese sent Ambassador Saburo Kurusu to Washington, the president penciled a note to his secretary of state, Cordell Hull, that he was willing to resume economic relations—including selling oil—within six months providing Japan did not expand farther into Indochina and did not invoke the Tripartite Pact, joining Germany, should the United States become involved in the war.

Michael Barnhart

I'm not sure FDR was the main push for sanctions, but the asset freeze (which became an oil embargo) was critical in deciding a long debate between Japan's army and navy as to whether to attack south (against British colonies) or north (against the USSR) in 1941. There was oil south, so the navy won this debate (with the proviso that after the Dutch East Indies were secured, the army then could attack the Soviets if the moment seemed ripe). The embargo cut that Gordian knot. Could the United States have bought more time by avoiding the embargo? Unquestionably, but what if the army had triumphed after the constitutional crisis and attacked the USSR in late 1941? I'm not certain the outcome would have been favorable to U.S. interests. Waldo Heinrichs' *Threshold of War* makes such a case at length and persuasively.

Collision in the Pacific

Believing they had to expand to survive, the Japanese sought to establish the Greater East Asia Co-Prosperity Sphere, a new order in the Far East that would exclude American and European imperialists.[10] On the other hand, America was committed to resisting Japanese aggression and neither side wanted to back down on China and Indochina.[11]

Q: *Was war between the United States and Japan inevitable?*

Douglas V. Smith

The purpose of the Strategy and Policy course here at the Naval War College is to educate officers to make better decisions in senior command and staff positions. To do that, students are asked to critically evaluate the strategies used by combatants on both sides in wars and major conflicts, with an eye to identifying alternative strategies open to them that might have served better to achieve the political objectives of their governments. Thus, this is one of our favorite questions for our students.

Certainly, there was an "America First" or "Fortress America" sentiment prevalent in the American population and to a lesser extent in the Congress, based, perhaps, on our experience in World War I. It is entirely possible that the Japanese could have attacked the Dutch holdings in Southeast Asia, and even possibly the British holdings there as well, without bringing the United States into the war against them. Thus, they would have gained access to the oil and other resources critical to their army operations in Manchuria and China without having to contend with the American industrial giant.

The Japanese—in much the same way that they patterned their navy after the British model and brought in German advisors for their army, replacing the French, because these were perceived as the most capable in the world—patterned their government on the Prussian system during the Bismarckian period. Thus, they required, by law, a serving member of the army in their cabinet. If the officer assigned dissented and withdrew from the cabinet, a new cabinet would have to be formed. But the requirement for a serving army officer still stood. The same, or a like officer, would be part of the next and any other cabinet formed. This provision of the constitution gave the army ascendancy over Japanese policy and strategy over time—and the Imperial Japanese Army needed oil. The navy's job, secured as their vital mission, was to ensure the free access to and transportation of oil. With the Japanese economy on the verge of collapse and with the U.S. Navy astride Japan's lines of communication (LOCs) with the required oil in Southeast Asia, it is extremely doubtful that the Japanese governmental or military leadership would have taken the chance of leaving their LOCs vulnerable when to do so could court disaster.

H. P. Willmott

There was a certain logic in terms of Japan fighting the United States. There was the old problem of better to fight the external enemy than a civil war, and there was the logic of initiating a war before the United States had grown into its full strength. There was also the calculation that Germany would prevail within Europe and therefore present the United States with an obvious distraction. There was also a certain logic to the Japanese decision to strike against the United States in 1941, and in one respect we must be careful in not reading the record backward. The Japanese leadership did not understand the United States and it most certainly never understood the economic and industrial capacity of the United States. No one, not just the Japanese high command, could grasp the fact that the United States could seriously plan to build 100,000 aircraft in a single year. The world became familiar, after 1941, with the reality of American industrial stamina but the scale of American production—at its peak one aircraft every 254 seconds, 140 Liberty Ships in a month, foreign aid equivalent to the sum needed to raise two thousand infantry divisions—was simply unthinkable in summer 1941. Now it is possible to argue that the real point here was that a nation ruled by a god and watched over by the gods could not be defeated, and one would not belittle such consideration in any way, but the basic point was that the Japanese high command, in initiating a war with the United States, began a war with a country that had two navies. The Imperial Navy fought the U.S. Navy to a standstill in the first year of the Pacific War, but its problem was that it fought itself to a standstill at the same time, and by November 1943 the United States had a second Navy, a Navy largely built over the previous two years, coming online, as well as a third fleet destined to appear in 1944. It was this second Navy, a wartime creation, that after November 1943 took the tide of war from the Gilbert and Ellice islands via the Marshalls and Marianas across the Pacific, and it did so on the basis of an overwhelming advantage of numbers. What happened after November 1943 was that the U.S. Armed Forces developed the capability to subject an objective to attack by land-based aircraft, to move, to isolate the objective from outside support by the use of carrier formations, to subject the objective to amphibious assault, and then to repeat the process against the next objective. It was not particularly pretty or imaginative but it was remarkably effective, and such was the disparity of strength that, in the first nine months of 1944, Japanese aircraft, warships, submarines, and mines failed to account for the sinking of a single American fleet unit, submarines discounted.

This failure to understand the United States in terms of depth of resolve and industrial, financial, and economic wherewithal was fundamental to Japanese failure, but one would suggest that the rivalry between the Japanese armed services was very important in shaping decisions in 1941. The Imperial Navy sought a course that would ensure against any Imperial Army move against the Soviet Union. Prima facie, this would seem preposterous: the Imperial Army had saddled itself with an unwinnable and disastrous war in China and had been hammered in 1938 and 1939 in border clashes with Soviet formations. However, once the German invasion of the Soviet Union began, and most certainly in the first three months of operations when the Japanese high command was making decisions, there seemed no alternative to speedy, complete, and comprehensive German victory: there was genuine incentive to move against the Soviet Union. But this division between the two services was far more fundamental than mere policy differences would suggest. Factories producing aircraft had to set up separate assembly lines for army and navy aircraft, and Japan never completed an atomic bomb project for the very simple reason that it had two, one for the army and the other for the navy. Japan was so disorganized that it did something that would seem to be impossible: it made the Confederacy look well organized. This comparison is not a simple throwaway line, and for a reason that is seldom considered: the comparisons between the Pacific War and the American Civil War are very close indeed. The Union advantage over the Confederacy in terms of population, communications, industry, and sea power were very similar to those of the United States relative to Japan, and in both conflicts the Union and the United States waged war on the basis of mass, firepower, and shock action. The battles fought in Virginia have their parallel in the southwest Pacific; the drive across the Pacific has its parallel in the drive down the Mississippi, with the enemy split in both conflicts; and finally, in 1945 there were the naval and air efforts directed against the Japanese heartland that compare to Sherman's march through Georgia into the Carolinas. Both wars lasted about four years, which, historically, is short. It is also worth recalling that wars are very seldom decided at sea. Armies decide wars on land, but the Pacific War was very unusual in that it was a war primarily decided at sea—though its joint dimension and its economic and political aspects must be properly acknowledged—and it was decided in what was a very short period of time.

Frank Shirer

Yes, war with the United States was inevitable, based upon the Japanese not ending their war against China. The United States had militarily supported China dating to back to when the Japanese invaded Manchuria and established the puppet-state of Manchukuo. That support, both official and popular, increased when the Japanese invaded China proper and news of Japanese atrocities such as the Rape of Nanking were reported in the news and in the movie news clips. The Japanese refusal to end their war with China was why President Roosevelt declared an embargo on scrap iron and other strategic materials in 1940. When this embargo showed no effect on Japanese foreign policy, he declared an oil embargo in mid-1941 and quietly authorized the recruitment of American pilots, known as the American Volunteer Group (a.k.a. the Flying Tigers). President Roosevelt refused to agree to the Japanese demands to stop the United States' support for China and end the economic embargo. The Western oil embargo was the key element in the Japanese decision to strike south and seize the Dutch East Indies, because Japanese oil reserves would have been depleted by late 1942. There were no oil fields in Siberia.

Harold Goldberg

Once we accept the inevitability of war, it seems that we have to accept the inevitability of the attack on Pearl Harbor as well. Framing some of this discussion in terms of the lessons of the Russo-Japanese War, we have a smaller power deciding that war with a much larger adversary cannot be avoided. Once that calculation is made, the smaller nation has an incentive to begin the war on its own terms and its own timing. Japan in 1904 and in 1941 recognized that time was not on its side if war broke out a year or two down the road.

In 1904, this analysis led to the surprise attack on Port Arthur. Although Admiral Togo Heihachiro, commander in chief of the Japanese fleet, did not achieve total success in February 1904, the Japanese navy's raid, combined with subsequent naval skirmishes, gave Japan control of the sea throughout the duration of the war with Russia.

The Russo-Japanese War was further determined by one large and decisive naval battle at Tsushima in 1905. Yamamoto, commander of the Japanese navy in 1941, fought in the Battle of Tsushima (where he lost two fingers). Certainly, we can see a pattern in a surprise attack followed by one gigantic naval battle (1904–5) as we move into the Pacific War. While the Russo-Japanese War would not be the only factor in Yamamoto's thinking in 1941, it seems reasonable that Yamamoto thought about an attack on the American fleet docked at Pearl Harbor followed by an attempt to finish the U.S. Navy by enticing it to Midway for a Tsushima-like confrontation. Not surprisingly, the Japanese attack fleet at Pearl Harbor was flying pennants from Admiral Togo Heihachiro's battleship *Mikasa*.

Summary

Perhaps war was inevitable.

Japan viewed America's two-ocean Navy and the relocation of the U.S. Pacific Fleet to Hawaii as physical threats, while the oil embargo and freezing of Japanese assets amounted to economic blackmail. Lacking resources, Japan felt she had to expand or forget about being a major player on the world scene.

For America, Japanese aggression in China and French Indochina was unacceptable, as was the threat posed against U.S., British, and Dutch interests.

The Pacific Ocean was huge but not big enough for two nations headed for a nasty showdown. By late 1941, the only questions that remained were who would strike the first blow, where it would fall, and how effective it might be.

Two

PEACE OR WAR: 1941

AS DIPLOMATS MOUNTED a last-ditch effort to avert hostilities, Japan's war machine geared up for operations in Borneo, Java, and Sumatra.[1]

Japan's plan was ambitious: strike with surprise at multiple targets, including Singapore, Burma, the Dutch East Indies, the Philippines, and Hawaii. If successful, the Japanese would gain key resources, like oil, tin, and rubber, and would extend their empire across much of the Pacific.[2]

Time to Strike

In January 1941, Admiral Isoroku Yamamoto, commander in chief of the Japanese Combined Fleet, called for a feasibility study of an air attack on the U.S. Pacific Fleet at Pearl Harbor. Yamamoto discovered that there was a reasonable chance of success, if all six fleet carriers of the First Air Fleet participated and if the plan remained a secret.[3]

It would take time to put everything in place. Specially modified aerial torpedoes for Pearl Harbor's shallow waters would not be available until late October, and Yamamoto's newest carriers, *Shokaku* and *Zuikaku*, would not be ready until November.[4]

In early November, the Navy General Staff issued operation orders for the war, including the surprise assault on Oahu.[5] Emperor Hirohito approved the plans on December 1, just six days before the raid.[6]

Q: *Could Japan have attacked the United States earlier than December 1941?*

Stephen D. Regan

Japan could not have nor would it have attacked Pearl Harbor prior to December 1941. The governmental unrest within Japan clearly indicates that the extremist Imperial Army was willing to do just about anything to expand Japan's influence in Asia, but it did not control large sections of the government, was openly opposed by the Imperial Japanese Navy, and had not convinced the emperor of the necessity for an Oriental version of Manifest Destiny. It was late into 1941 that General Hideki Tojo, the minister of war, had wrested enough power to convince his emperor to entrust Japan to his hands. Tojo still had to fight a political battle with the IJN. Navy minister Koshiro Oikawa, foreign minister Admiral Teijiro Toyoda, and their stalwart allies, including Admiral Isoroku Yamamoto, faced increased humiliation from the army and Tojo. Navy leaders, openly accused of cowardice, feared that the army would assassinate Yamamoto and others if they did not acquiesce to the army's demands.

The December 7 date seems about as early as the Japanese navy could pull off an attack. Yamamoto worked as quickly as he could in developing a plan, training his pilots and crews and organizing his fleet. No political or military evidence exists to suggest that any earlier attack could have been possible.

A bigger question may be what would have been Japan's reaction had it waited for further negotiations with the United States. Tojo's desire for war might have been severely tempered if Washington had agreed to some semblance of placating the economic sanctions by the summer of 1942. Tojo's opposition may have been able to gain an upper hand in swaying the actions of the nation.

Earlier than December 7? Not a chance. Later than that? That is a question to be pondered. This suggests that if Japan would have shown more patience, it may have avoided war with the United States entirely or at least might have been in a position to negotiate peace.

Jon Parshall

In my opinion, no, at least not successfully. The ability to attack Pearl Harbor was predicated on the ability to mass and coordinate carrier air power. However, as of early 1941, all the world's major fleets were still using carriers in ones and twos. The Imperial Navy had been the first to make the conceptual leap necessary to mass their aircraft carriers into a cohesive fighting force. That leap resulted in the creation of the First Air Fleet in April 1941. First Air Fleet rightly deserves to be seen as a revolutionary development, as important as the creation of, say, steam-powered warships, or the launching of the *Dreadnought*. For the first time in history, a naval air fleet existed that had the potential to deliver strategically meaningful quantities of firepower to the battlefield. It marked a decisive move away from viewing carriers as primarily scouting and tactical raiding assets, and toward their usage as a primary, and strategically important, arm of battle.

However, it's one thing to create an administrative entity called First Air Fleet; it's quite another to create a cohesive fighting force that can carry out its intended mission of delivering concentrated air power. It has to be recalled that in 1941 there was no body of carrier doctrine anywhere on the planet that talked about how one should go about accomplishing that task. The Japanese had to invent theirs from scratch. And that took time, and many operational exercises, as the Japanese feverishly hammered out the nuts and bolts of how these ships would operate together. How far apart should the carriers steam? How should aircraft be launched and formed up? How should the attack formations be organized? Should the aircraft formations be composed of separate groups of planes from each individual carrier, or should they be combined and then organized by type? Who should command them? How many aircraft should be launched from each carrier when carrying out a strike? How should the fleet be defended from enemy air attack? All of these questions had to be answered. Recent research from other experts in this field has shown that the Japanese were still evolving their doctrine on these very fundamental issues up until practically the time Nagumo sailed from Hitokappu Bay. So, in my mind at least, the organization and doctrinal basis for Pearl Harbor simply didn't exist until sometime in the latter third of 1941.

The other question revolves around that of available force. At the time First Air Fleet was formed, the Imperial Navy only had four fleet carriers in the inventory—*Akagi*, *Kaga*, *Hiryu*, and *Soryu*. The large carriers *Shokaku* and *Zuikaku* would not be added to the fleet until September/October 1941. If you look at the Pearl Harbor operation, it's clear that the Japanese were doing more than attacking a naval anchorage. They were also attacking and suppressing the American air power (both Navy and Army Air Force) on Oahu. To do that, you have to have sufficient assets to attack not only the anchorage, but also the major airfields on the island of Oahu—Hickam, Kaneohe, Wheeler, and so forth. That requires a lot of aircraft. The war games that the Japanese ran preparatory to the attack brought home this fundamental point—if you didn't attack with sufficient force, the mission would be a failure. Therefore, it was vital that Carrier Division 5 (*Shokaku* and *Zuikaku*) be a part of the attacking force. That, again, strongly implies that a successful attack against Pearl Harbor could not have been undertaken until late in 1941.

North or South?

Could Japan have picked out another opponent?

Some in Japan wanted to fight the Soviet Union, at least until the summer of 1939, when the Soviets soundly thrashed Japanese forces near Nomonhan in Mongolia. The Battle of Khalkhin Gol revealed that the Soviets were superior to the Japanese army in several categories, including armor and artillery. As Japan's respect for Soviet capabilities grew, support waned for further expansion near Korea, Manchuria, and the Soviet border. Soon thereafter, Hitler concluded a nonaggression pact with the Soviets.[7]

In April 1941, Japan signed her own nonaggression treaty with the Soviet Union, allowing the Soviet troops on the Manchurian border to rush to the defense of Moscow following Hitler's invasion of the Soviet Union in June 1941.[8]

Q: ***If the United States had done more to improve relations with Japan, would the Japanese have attacked the Soviet Union instead? What if the Japanese had not signed a nonaggression pact with the Soviets and had maintained troops near the Soviet border in June 1941?***

Robert Mrazek

I believe that World War II could well have ended very differently if Germany and Japan had acted as true allies, and had together executed a strategic plan that took into account their mutual strengths in first defeating the Soviet Union and Great Britain before eventually confronting the United States. If Hitler had confided his plan to attack Russia in the spring of 1941 to his Japanese ally, and had secured from the Japanese their commitment to attack the

Soviet army in Manchuria in a two-front strategy (perhaps in exchange for a share of the oil in the Caucasus), it is hard to conceive of any other result than an Axis victory over the Soviet Union. In October 1941, Soviet leader Joseph Stalin was only able to stop the spearhead of the German army from capturing Moscow by bringing his troops from eastern Russia that were deployed there against the Japanese. If the Japanese had joined Germany in attacking the Soviet Union in the fall of 1941, Stalin would have had to retreat into Siberia. There would have been no Murmansk convoys. The Soviet army would have been destroyed as a unified fighting force.

I strongly doubt whether the American people would have been ready to go to war over the defeat of the Soviet Union. Isolationism was still rampant, and more than three-quarters of Americans were opposed to joining the war. Hitler would have now been free to turn all his might on England. Again, a combined strategy by Germany and Japan could have accomplished this goal.

In the Pacific, the Japanese could have launched the same attack against the Western colonial powers that they initiated in December 1941 with one important exception. Instead of attacking the U.S. forces in the Philippines and Hawaii, Japan should have avoided a direct confrontation with America and focused its efforts on all the other Western colonial powers, principally England. Thus, they would have conquered oil-rich Indonesia, Java, Sumatra, Hong Kong, Singapore, Malaya, and Burma. They would have secured the precious oil they needed, and acquired an empire of 150 million subjects.

Concurrently, the Germans could have invaded England without direct involvement from the United States, aside from its merchant marine convoys. It is hard to imagine President Roosevelt convincing the American people that they needed to go to war to protect England or its empire, certainly not with the unbridled zeal that resulted from the Japanese attack on Pearl Harbor.

With England and its powerful navy out of the war, nothing could have stopped the Axis partners from securing the Mediterranean, North Africa, and the Middle East. With Germany and Italy attacking from the west, and Japan from the east, India would have fallen easily. The Axis armies could well have met in the Middle East to divide the wealth of oil-rich Arabia.

If the Japanese and Germans had avoided the blunder of either attacking or declaring war on the United States in 1941, if they had focused their combined might on defeating their other adversaries first, America could well have faced a challenge of far greater magnitude in 1942.

Harold Goldberg

This excellent hypothetical question deserves a hypothetical response that examines the "what if" aspect of the problem. Let's construct a scenario for improved relations that does not assume any modification in behavior on the Japanese side (since that was most unlikely): all of the theoretical changes in policy would have been made by the United States. After examining the conditions necessary to create a climate of improved relations between the United States and Japan, we might be able to determine whether events could have moved in a different direction.

We would start with President Roosevelt choosing Joseph Grew, then American ambassador to Japan, as secretary of state, rather than Cordell Hull. That selection would have been a response to the Soviet choice of A. Troyanovsky, their ambassador to Japan, as ambassador to the United States. As a result of Grew's influence, the United States would have muted its criticism of Japanese aggression in Manchuria and accepted the Japanese claim that they were filling a power void left by a weak Chinese government.

Let's also agree that at the end of the 1930s, the United States would not have moved Pacific Fleet headquarters from California to Hawaii (1940) and would not have passed the Two-Ocean Naval Expansion Act. Japan therefore continued on its aggressive course without concern for American reaction.

Imagine that the United States treated Japan's occupation of all of Indochina in 1941 as it had treated the invasion of China—in other words, trade went on without an American embargo. As a result, Japan continued to import oil and other strategic goods from the United States and did not feel the need to move south toward the Dutch East Indies and the Philippines.

In Japan, changes in American behavior might have had a dramatic impact. Without pressure from the Japanese military for a decision on war, Prince Konoe rather than General Tojo would have remained as premier of Japan in the fall of 1941. With American goods flowing to Japan, the Japanese leadership would have decided to finish off its traditional enemy—Russia—and would have moved out of Manchuria to join Germany's war in the USSR in the late summer and fall of that same year. As a result of Japan's decision, the Soviet Siberian army would not have been available for the Battle of Moscow in December 1941. Pushing this scenario a bit further, the Germans would have taken Moscow, and then arrested and executed those leaders of the Soviet government who had not been evacuated. Let's assume that Stalin would have been captured and shot, or at the very least removed from his position while the USSR disappeared from the map. For Japan, there was no need for Pearl Harbor. The USSR fell; Britain negotiated a deal with Germany on the division of Europe; the Axis won the war.

Having made this argument, we can now destroy it. The political and military realities of the time largely negated the possibility that the above changes could have occurred. The imagined scenario is flawed because it separates U.S.-Japanese relations from U.S.-Chinese relations, Japanese-Chinese relations, Japanese-Soviet relations, and Japanese-German relations.

In other words, it was not possible for the United States to improve relations with Japan without any change in Japanese foreign policy—our original assumption.

The United States could not abandon the Nationalist government of China and accept a decade of Japanese aggression in Manchuria, China, and Southeast Asia without something in return from Japan. Cordell Hull was not the only actor in shaping American foreign policy and Grew by himself could not have engineered a complete reversal of policy.

Despite American caution (until 1940–41), it was unlikely that the American government would have shifted policies dramatically away from China and toward Japan. Roosevelt had to act incrementally while awaiting shifts in American public opinion, but a protracted acceptance of Japan's unchecked aggression was not likely.

For the scenario to be realistic Japanese foreign policy had to change; unfortunately Japan was not motivated to change. Only military defeat would lead to a decrease in military influence on political appointments and a significant reorganization of the Japanese government and cabinet. That dramatic realignment of forces in Japan occurred in 1945, after the military was discredited and a new political structure was imposed on Japan from the outside.

Even if we accept something like the scenario that presumed a U.S. acceptance of Japanese expansion followed by the fall of the Soviet Union, the logic of events would have pushed Japan toward war with the United States. Flush from victory in Russia, the Japanese military would have continued on their path of aggression and eventually turned south.

Japan would have revisited its designs on the Dutch East Indies, with the Philippines standing in the way of further Japanese expansion. Japan would have had to target the American fleet, even if the date were December 1942 rather than December 1941.

The story ends right where it started.

Bypassing the Americans

Although the Japanese coveted the Dutch East Indies, the Malay Peninsula, Burma, and Singapore, was it necessary for them to attack American interests like the Philippines?

Several members of the NGS, including Admiral Osami Nagano, felt that Roosevelt would have a tough time lining up support for a war to bail out the British and Dutch. However, Yamamoto was not about to rely on the United States sitting on the sidelines while Japan made her move in Southeast Asia, not when the Americans could threaten his eastern flank.[9] To protect the shipping lines with the East Indies, and to capture the agricultural wealth of the Philippines, Japan opted to invade the American-controlled archipelago.[10]

Q: *Was it feasible for Japan to bypass U.S. possessions, including the Philippines, while carrying out operations in December 1941?*

Jon Parshall

From a political perspective, it's clear that bypassing the Philippines would have offered the Japanese important tangible benefits, and would have made Roosevelt's life very difficult in terms of trying to get a declaration of war out of the U.S. Congress, even if the Japanese had attacked the British and the Dutch. After all, the Germans had attacked the British and the Dutch, and we hadn't declared war on Germany, despite the fact that they were demonstrably the more dangerous of the two potential opponents in any upcoming war. I think the odds that FDR would have been able to get Congress to act immediately if the Japanese had attacked, say, Singapore weren't good.

Fortunately for Roosevelt, though, I also can't conceive of the Japanese being imaginative enough to take advantage of these political dimensions when conceptualizing their Phase I military operations in the area. That may sound crazy; after all, this is the nation that came up with the attack on Pearl Harbor, which was pretty darned unconventional. But I don't think any trained military man in Japan would have allowed a potential threat to the supply lines to the Southern Resource Area to sit unreduced. They would have wanted to eliminate those threats as soon as possible.

The natural resources in places like Java and Borneo, which were the basis of the war in the first place, had to be brought home to the Home Islands for processing, manufacture, and distribution. The seaborne lines of transportation from those areas ran straight through the South China Sea. They were exposed to direct attack by air and sea assets operating out of the Philippines for thousands of miles of their length. The Philippines were ideally situated to pose a major threat in this respect.

Bear in mind, too, that at this point in time no one knew that the B-17 wasn't effective against shipping. So the fact that a relatively large contingent of B-17s was sitting on Clark Field represented a potentially dire threat to Japanese convoys. The same was potentially true for submarines operating out of Subic Bay. The Japanese didn't know our torpedoes were defective (and neither did we). Granted, the overall American military presence in the Philippines wasn't all that large, but had the archipelago been left unreduced, we could have reinforced our presence there substantially, making an eventual Japanese attack even more difficult to pull off. Given that scenario, the Japanese army would have wanted to take the place as rapidly as possible.

The other thing to remember is that the military (mostly the army) was running foreign relations and diplomacy in Japan. And the Imperial Army was clearly out of its element in this respect, or they just plain didn't care. Pearl Harbor is a case in point: meticulous attention to the military details, accompanied by ham-handed attention to the diplomatic niceties, as well as a complete misunderstanding of what the attack would do to galvanize America's political will to wage total, unrestricted warfare against the aggressor. Pearl Harbor was a spectacular operational success, but a diplomatic and political catastrophe for the Japanese. In fact, you can make a credible argument that the Japanese essentially lost the war as soon as the first bomb landed on the seaplane ramp on Ford Island. Is this the sort of military capable of subordinating perceived operational imperatives in order to bolster the diplomatic dimensions of the war they want to wage? I don't think so.

Stephen D. Regan

The Japanese feared U.S. involvement in their desire to control all of Asia. They believed that America would indeed enter the war in the Pacific on behalf of the British, French, and Dutch, whose colonies were Japan's goals. Yamamoto was opposed to an American war but, once convinced that a means to set back American involvement in the Pacific was the only answer to Japan's needs, he quickly developed the Pearl Harbor attack.

Japan evidently did not appreciate the situation in Washington. Roosevelt openly recognized that England had to be preserved and that a Nazi Europe was unacceptable vis-à-vis American interests, especially industrial and economic interests. Japanese intelligence failed to comprehend that the Pacific was of secondary interest to the United States. Secondly, Hawaii and the Philippines were the limits of our overall concern. Without a direct attack on either of these two areas, FDR would have had a massive opposition to involvement in the East.

The Philippine Islands had been an American protectorate since the Spanish-American War; however, the impetus for an American empire died after World War I. While Douglas MacArthur's reign as head of the Filipino army ensured some Washington interest in the islands, all data since 1940 seem to suggest that Roosevelt and the American citizenry were only moderately interested in the islands. Bluntly, neither Republican nor Democrat political figures showed much interest in going to war over the Philippines or Southeast Asia.

Had the Japanese invaded the Dutch East Indies, they would have faced censure and diplomatic condemnation. Further economic sanctions would have been mandated and the diplomatic community would have been aflutter over such military expansion. Little suggests that any neutral nation would have declared war on the Japanese. It seems rather probable that the Japanese Empire could have expanded all throughout French Indochina and Southeast Asia with little more than nasty letters exchanged.

It is possible that the fleet in Hawaii may have been sent on maneuvers in the general areas of the South China Sea and along the east coast of Japan. But more offensive behavior is unlikely. Washington was hoping to avoid conflict on a two-ocean front. Roosevelt knew that someday he would have to deal directly with the Japanese problem but until the conflict in Europe was won, the Pacific was on the back burner.

We must remember that the U.S. Navy and many politicians were adamantly opposed to moving the fleet from California to Hawaii. Roosevelt ignored their concerns because he wanted to posture a little to the Japanese, just like his cousin Teddy Roosevelt did with the Great White Fleet. Clearly, he did little more than send the ships to operate out of Pearl Harbor instead of California. No major buildup was apparent.

Even an invasion of the Philippines is questionable in terms of an American declaration of war. Isolationism was still the mind-set of most Americans, and it was this obdurate attitude toward Europe that occupied the myriad of minds in the White House. FDR had not roused the general public against Hitler, whom he saw as the real threat to Western ideals. An attack on the Philippines would have been deemed unimportant by the Republicans, the conservative Midwest voters, and the parents of eighteen-year-olds. Roosevelt could not have nor would have taken a position of unconditional surrender because of an assault on the Philippines.

Had Japan merely moved into the remainder of Asia without directly attacking the United States, the American war in the Pacific would not have happened until well after the fall of Germany. Russia, on the other hand, may have seen Japan's southern movements as an excuse to gain some territory north of the Japanese islands and in Korea, to say nothing about keeping China at bay. Conjecture about another Russo-Japanese war is open for discussion.

Jim Hallas

From a military standpoint, it was unacceptable for the Japanese to seize the Dutch East Indies while leaving the American bastion in the Philippines unmolested. The Philippines lay directly across Japanese supply lines to the East Indies, leaving those routes vulnerable to interdiction at any time. There was no point in seizing East Indies oil if that oil could not be safely moved. Japanese strategists were acutely aware of this threat. Control of the Philippines was considered so crucial that initial Japanese war plans called for moving against the East Indies only after the Philippines had been completely subdued.

Still, leaving the military necessity/reality behind for a moment, it is interesting to speculate on what the U.S. reaction might have been had the Japanese gambled on simply seizing the oil fields in the Dutch East Indies and ignored the Philippines and other American possessions. On December 1, 1941, President Roosevelt assured the British ambassador that if the Japanese attacked

British possessions or the Dutch East Indies, "we should all be in this together" and the Dutch and British could count on "armed support."

Of course, such a commitment was not entirely up to Roosevelt. War would need to be authorized by Congress, and that authorization—and the acquiescence of the American public—was by no means certain. After all, the United States had stood by while Nazi Germany occupied most of Europe. Would the American people have been prepared to go to war over a distant Dutch colony? It seems unlikely.

Faced with seizure of the East Indies alone, it is conceivable that the U.S. government would have talked and blustered but failed to act, at least for the moment. For years, the U.S. government had done little but complain regarding Japanese actions in China, but the American mood was hardening. In late 1940, the United States imposed bans on the export of critical materials such as scrap steel to Japan. When Japan seized southern Indochina in 1941, FDR froze Japanese assets in the United States. A reckoning of some sort was clearly only a matter of time.

In the end, war would only have been deferred, not averted, by bypassing U.S. possessions. Access to oil would have eased the mounting pressure on the Japanese to go to war before their navy was rendered impotent due to lack of fuel, but it would not have satisfied Japanese ambitions. In light of Japanese plans for the Greater East Asia Co-Prosperity Sphere, the Philippines would have been targeted eventually; in any case, the United States was prepared to fight to maintain its authority in the Pacific. The attack on Pearl Harbor ensured that the nation entered that fight immediately and with a will.

Summary

In 1941, Japan had options beyond attacking America, like working with Nazi Germany to take down the Soviets, but the lure of Dutch East Indies oil proved to be irresistible.

Could Japan have bypassed U.S. possessions like the Philippines while invading the Dutch East Indies? Maybe, but many in Japan, including Yamamoto, decided that war with America was inevitable. Better to attack Hawaii at the outset rather than allowing an intact U.S. Pacific Fleet to sail against Japan's newly won territories.

With the die cast, Yamamoto prepared to send the Imperial Japanese Navy across the Pacific, hoping to devastate Pearl Harbor to give Japan the breathing space it needed to consolidate its conquests.

Three

DECEMBER 7: PEARL HARBOR

JAPAN NEEDED TIME—time to overcome the Philippines, Burma, and the Dutch East Indies, time to traverse the Malay Peninsula to capture the British fortress of Singapore and its all-important naval base.

In Yamamoto's mind, he had no choice: he had to obliterate the U.S. Pacific Fleet at Pearl Harbor.[1] While some on the Navy General Staff felt the raid was too risky, Yamamoto threatened to resign if the operation did not move forward, effectively silencing the opposition.[2] After reviewing coded reports on the U.S. Pacific Fleet sent by the Japanese consul general in Honolulu, Yamamoto selected Sunday, December 7, as the strike date.[3]

Warnings

The IJN's attack on Oahu by the six carriers of the carrier fleet, Kido Butai, carried out in two waves of more than 350 torpedo planes, Zero fighters, and bombers, caught the Americans totally off guard. Eight U.S. battlewagons were present that Sunday morning, including six on Battleship Row on the eastern end of Ford Island, in the middle of Pearl Harbor.

The first wave struck at 7:49 a.m., the second wave about an hour later. In the end, of the battleships, *Arizona* and *Oklahoma* lay at the bottom of the harbor, *West Virginia* and *California* were sunk but salvageable, *Nevada* was beached, and *Tennessee*, *Maryland*, and *Pennsylvania* were damaged. Several other ships were sunk or damaged, approximately 340 planes were damaged or destroyed, and more than 2,400 Americans lost their lives, including more than one thousand on the *Arizona*.[4]

There may have been signs of the impending attack for those able to discern them. Thanks to code-breaking, the Americans knew that Tokyo had ordered its consulate in Honolulu to relay detailed information on warships at Pearl.[5] In fact, on November 15, Tokyo called for twice-weekly reports on the exact location of U.S. ships.[6]

Then, ten days before the attack, overseas commanders received a war warning from the chief of naval operations, Admiral H. R. Stark, specifying the Philippines, Borneo, and the Thai and Kra Peninsula as potential targets, with no mention of Hawaii.[7] Washington's final warning to the Army and Navy commanders in Hawaii, General Walter C. Short and Admiral Husband E. Kimmel respectively, sent out less than an hour before the first bombs began to fall, reached Short and Kimmel *after* the Japanese attack.[8]

In the hours leading up to the raid, the destroyer *Ward* sank a Japanese midget sub near the harbor entrance. Army privates operating a radar station on the northern tip of Oahu also sent in reports of a large blip of planes. Unfortunately, the officer in charge of the Army Aircraft Warning Center at Fort Shafter, Lieutenant Kermit Tyler, assured himself a spot in the history books by assuming the planes were B-17s from the mainland and telling the privates not to worry about it.[9]

Q: *What if the Americans had heeded potential warnings in advance of the Pearl Harbor raid?*

Donald Goldstein

If the U.S. forces at Pearl Harbor had received just a few hours' notice of the impending Japanese attack, and had acted on those warnings, it might have been an even greater disaster for America. If the U.S. ships had sailed and left Pearl Harbor, a number of ships might have been sunk at sea. The death toll might have been even greater. Admiral Chester Nimitz had said that it would have been catastrophic to have lost the American ships at sea, away from Pearl Harbor. The casualties would have increased and the ships that were sunk would not have been raised. Most of the ships sunk in the shallow waters of Pearl Harbor were raised to fight another day. You also have to wonder how capable the U.S. Navy was early in the war. If the U.S. Pacific Fleet had been able to come to blows with Nagumo's carriers on December 7, would the Americans have had the experience and the ability to inflict severe damage on the Japanese ships?

Jon Parshall

The conventional wisdom on this matter—that a little warning for the Americans would likely have resulted in a greater catastrophe for the American battle line—is, I think, right on the money. I had the pleasure of war-gaming this for a Discovery Channel television show a few years ago, playing the Japanese side. If you give the Japanese the opportunity to launch a coordinated strike against the American battle line at sea, the results are most likely going to be very, very bad for the U.S. Navy. U.S. ship-borne antiaircraft defenses were very weak at this stage of the war, and the older battleships were slow-moving targets. Without air cover from Oahu (which would have been difficult to coordinate), I think the Americans would have been facing a catastrophe of Tsushima-like dimensions, with potentially dozens of ships lost, and thousands more sailors killed than were lost during the actual attack.

Keith Allen

With a few days' warning, a great deal could have been done—strong air and submarine patrols; concentration of the fleet off Hawaii, including the carriers; defensive preparations on Oahu itself, including standing fighter patrols, dispersal of aircraft, full manning of antiaircraft guns, and so forth.

There is a school of thought that holds that we were lucky that the fleet was attacked in port, because most of the battleships could be salvaged, and that we would have been worse off if our ships had been able to put to sea and had been attacked there. In my opinion, this is wrong (and I also think there is a certain element of sour grapes in some such efforts to minimize the Japanese tactical success on December 7). This theory seemingly assumes that the Japanese would have scored as many hits on a fleet maneuvering at sea as they did on one moored in Pearl Harbor, which makes no sense; a maneuvering target is much harder to hit than a stationary one, and the fleet's defensive fire would have been much more effective if it had been fully alerted.

With three to four hours' notice (or more realistically, one to two hours' notice, if the radar contacts had been heeded and the significance of the *Ward* contacts recognized), most of the fleet would still have been in port, but the guns could have been manned and the maximum condition of watertight integrity (Condition Zed) set. Our ships would still have taken considerable punishment but less than they actually did. More fighters could have been launched and inflicted higher losses on the Japanese.

Stephen D. Regan

The entire focus of the war would have been greatly altered had the United States fully appreciated the potential for Japan to attack Pearl Harbor. In retrospect, the United States had some cognitive awareness of Japan's ability to surprise Pearl Harbor with an aerial attack, as had been accomplished in war game and fleet exercises. However, viscerally the Americans did not believe that Japan could pull off such an audacious maneuver. Ethnocentrism played no small part in Navy and Washington political circles. Americans blatantly and outspokenly believed that the Asians were intellectually inferior, and that they possessed no parallel engineering, military, or manufacturing minds. Bluntly, the Japanese could not and would not attack Pearl Harbor and declare war on the United States, or so they thought.

Had naval authorities and political figures recognized the danger to Hawaii, and offered even a mere two hours' warning, the entire war would have been greatly different. The battleships would have had the chance to escape. With the big boomers saved, aviation would have taken a prescribed position as mere protection for the Big Guns. It is conceivable that the Japanese, finding no one home at Pearl Harbor, would have struck at the secondary targets, which they ignored on December 7, 1941: the dry docks, the oil storage facilities, and the submarine pens. Without oil, dry docks for repair, or submarine services the United States would have been unable to mount much of an offense for at least six months. No Coral Sea or Midway would have reached the history books.

Even if the Japanese had not successfully mauled the precious fuel and repair docks, the United States would have recognized the weakness in the Pacific and probably would have put the fleet on some sort of defense for at least a year, and the commander in chief of the U.S. Fleet would have focused on the Atlantic and Germany. The U.S. strategy against Japan would have been battleship-focused, and the internecine war between the Brown Shoe and Black Shoe (aviation versus battleship) departments would have been heated, mean-spirited, and politically deadly. If Yamamoto had continued his reliance on his carriers, the Navy would have seen her battleships and cruisers picked off one by one until President Roosevelt capitulated and signed some sort of negotiated peace settlement. The loss of the battleships was a huge blessing to the Navy and eliminated the internal philosophical squabbles. When all we had left were the carriers, then we focused on carrier warfare. With the big guns intact, the rivalry would have continued unabated.

The second aspect of a warning impact would have been the chance to get some planes airborne in time for some fierce fighter combat. With Japanese planes being involved in a melee and many being damaged, the coordinated attack would have failed. Without the coordination, Japan probably would have inflicted limited damage. Using Midway as a guide, uncoordinated attacks without the element of surprise would have been minimally successful. No doubt, many Japanese planes would have been shot down and many good pilots would have been lost. In the final analysis, an early warning may have placed the United States in a very difficult position, or it could have greatly altered Japan's offensive plans. Either way, the overall look of the war in the Pacific would have been greatly altered from what we experienced. No doubt, there would have been no Coral Sea, no Midway, and no Guadalcanal. It is worth noting that even with a full day's notice MacArthur failed to do anything to defend his station. His own ethnocentrism impaired his judgment, although he was quite knowledgeable about Japanese abilities. Evidently, Americans of all kinds were deep into denial.

Third Wave

While Japanese planes clobbered U.S. warships, several other attractive targets remained untouched, including Pearl Harbor's submarine base, power plant, oil tank farm, and ship repair facilities.

However, the commander of Kido Butai, Vice Admiral Chuichi Nagumo, ruled out additional attacks, citing concerns with antiaircraft fire, a dwindling fuel supply, and missing enemy aircraft carriers.[10] Nagumo could have searched for the flattops or launched another aerial assault on Pearl. Instead, he did neither, electing to retire, and Yamamoto refused to intervene.[11]

Without power, oil, or the ability to repair their ships, the Americans would have had a tough time maintaining a defensive posture in the Pacific, let alone launching offensive actions.[12] Admiral Chester Nimitz would later

Japanese photograph taken during the Pearl Harbor attack, showing Battleship Row as smoke rises from Hickam Field. However, Pearl Harbor's oil tanks were left untouched.

say that the Japanese had given the Americans a chance to catch their breath, while Admiral Kimmel claimed that the destruction of Pearl Harbor's oil tanks would have forced the U.S. Navy to withdraw to the West Coast.[13]

Q: *What if the Japanese had targeted oil tanks and ship-repair facilities in their second wave, or had launched a third wave to hit those targets?*

Hal Friedman

I find this question fascinating, in part because for some months now I have had the thought that Japan actually lost the war on December 7, 1941. By asserting this, I do not mean to revise the idea that Coral Sea–Midway-Guadalcanal was the turning point of the war, nor do I mean to argue that Japan's defeat was inevitable. In fact, my assertion does not even mean that had Japan taken different actions on December 7, it would have won the war. I'm not even sure of that. I am convinced, however, that if Japan had more thoroughly prosecuted the strike on Pearl Harbor, it would have made things infinitely more difficult for the United States.

Vice Admiral Chuichi Nagumo should have definitely listened to Rear Admiral Tamon Yamaguchi and Commanders Mitsuo Fuchida and Minoru Genda about launching additional strikes against Pearl Harbor. These officers, especially Fuchida and Genda, were convinced that additional strikes could have succeeded in knocking out Pearl Harbor as a functioning base for the United States, since the first two waves had succeeded so well in disabling U.S. air and naval power in Hawaii. As historians such as Gordon Prange have pointed out, not only was the Pacific Fleet's battle line disabled by the late morning of December 7, but half of the military's air assets on Oahu were destroyed by that time and most of the other half was damaged. Japan owned the air over Oahu and had local sea control. This probably would have continued even had the *Enterprise* and *Lexington* task forces shown up in the next day or two. The American carriers would have been outnumbered three to one and possibly worse, since the two carriers were not operating together. Clearly, the Japanese could have launched one additional strike, if not two, to destroy the Pacific Fleet's fuel and ammunition depots as well as strike at dry docks, hangars, and, perhaps even more importantly, the untouched Pacific Fleet submarine base.

With the destruction of these facilities, especially the dry dock and hangars, the United States probably would have had to pull back to San Diego and try to fight the Pacific War from an Alaska-California-Panama line. I cannot even begin to imagine how difficult it would have been to try to fight the naval or even air war from a Western hemispheric line. It might not have been successful for the United States in the long run. At the very least, it would have had to somehow reestablish Pearl Harbor as its main base in the Pacific Ocean area.

Nagumo was correct to worry about the safety of his carriers and keeping them safe in a long war. These were not his ships, but the emperor's. Twenty-two American submarines were docked at Pearl Harbor and had not been touched by the attacks. Two, if not three, American carriers were somewhere out there and needed to be accounted for. However, Nagumo's caution definitely demonstrated his lack of ability as a carrier air commander. The hasty plan Fuchida and Genda devised to strike Oahu again, draw in the U.S. carriers, and sink them did not fully take into account the refueling needs of the task force or lurking American submarines, but my guess is that the Japanese had "shock and awe" on their side for the rest of the day and perhaps longer. This advantage probably would have given them the time to complete the damage to Pearl Harbor and take on the U.S. carriers in the next day or two. With their tankers down south by that time, Japan could have completed turning Pearl Harbor into wreckage. I do not know if this would have given Japan the ability to invade and occupy Hawaii. As John Stephan and Jon Parshall have indicated, the Japanese had a particularly difficult time garrisoning their conquered possessions even while the war was going their way. Nor was the Fuchida-Genda plan without risk, but war is about taking risks and Japan had already taken its greatest one. Perhaps they should have gone all the way.

H. P. Willmott

The idea of an attack on the base facilities and oil storage depots on Hawaii simply takes no notice of two points: the Japanese carriers did not have the heavy bombs that would have been needed to have any chance of inflicting telling damage, and it is very difficult to see when the Japanese carriers could have conducted an attack. An attack could not have been conducted on the afternoon of December 7 because of fading light, and it could not have been conducted over the next couple of days because of fuel considerations, the need being for destroyers to get back to the oilers. Thus, one is left with an impossible problem of deciding when a follow-up attack could have been made.

Moreover, the real point is that even the most successful of attacks on base facilities and oil storage facilities would have availed the Japanese little. The speed with which the Americans put facilities in place suggests that Pearl Harbor would have been restored to fighting efficiency in a relatively short time—a year or so—while the surplus oilers and base facilities in California would have been sufficient to tide the U.S. Navy over even if every last gallon on Oahu had been destroyed. As long as the Americans continued to hold Oahu, whatever damage was inflicted would have been righted, and righted relatively quickly, with oilers sent forward to have available adequate storage in Pearl Harbor.

Jon Parshall

It would have been very bad for the Americans. However, the odds of this occurring were very low.

In fact, it is a common misconception that such an attack was even contemplated, or that it was a likely outcome of any follow-on strike. If you look at the targeting priorities for the operation that were given by Combined Fleet staff (i.e., Yamamoto), the list goes: 1) carriers, 2) battleships, 3) cruisers and other large combatants, 4) smaller combatants and submarines, 5) merchant shipping, and last of all, 6) logistics and repair facilities. Why this targeting priority? Japanese naval doctrine was ultra-Mahanian in its outlook. They believed fervently that destroying the major combatants of the enemy via decisive battle ipso facto led to sea control and victory in war. This, therefore, was the focus of any decisive naval operation. Logistics facilities simply did not fall within their area of interest in operational planning. Their viewpoint was that if you destroyed the enemy's warships, who cared about their logistics facilities? This same outlook informed every aspect of their navy: its doctrine, its battle training, its warship design, its mixture of warships, even its own lack of emphasis on things like amphibious warfare, gunfire support for ground troops, and (most tellingly) commerce protection.

If you look at that Pearl Harbor targeting list, it's clear that the initial morning attacks had barely chewed their way into item number three on the list. Indeed, there were still battleships left afloat and relatively undamaged (such as *Pennsylvania*). Given that, it's inconceivable to me that the Japanese would have jumped all the way down the list to item number six and said, "Let's go all out for those fuel tanks, boys!" It simply wasn't on their radar screen.

This myth, incidentally, comes directly from Mitsuo Fuchida, who stated in his 1953 book on Midway that he had mentally earmarked these logistics facilities for destruction as he was assessing the effects of the attack on the way back to *Akagi*. There is credible evidence to suggest that Fuchida created that fiction, though, as a result of hearing the incredulity from his postwar American interrogators that the Japanese had not attacked these facilities. Indeed, if you look at Fuchida's own United States Strategic Bombing Survey (USSBS) interview in 1946, he makes absolutely no mention of these matters. When asked why Pearl Harbor was not attacked again on December 7, he responded that their battle damage assessment was incomplete, and that the initial signs from the attack indicated that the American battle line had been destroyed, and that meant that they had accomplished their mission. End of story. Let's go home. Likewise, the supposed confrontation between Fuchida and Nagumo on the bridge of *Akagi*, with Fuchida supposedly begging Nagumo to attack these facilities, apparently never happened. Genda, at least, in his own account flatly denies that such an episode took place. All in all, Fuchida's account on this matter must be taken with a grain of salt (at the very least), and is in all likelihood a completely fabricated postwar fiction.

In summary, if the Japanese had attacked these facilities, it undoubtedly would have been very bad for the Americans. But the Japanese ability to conceive of these facilities as being strategically meaningful was extremely limited. As a result, the odds of the Japanese actually going after these installations were very low.

Get the Carriers

U.S. aircraft carriers were noticeably absent from Pearl Harbor on December 7. *Lexington* was on her way to Midway, *Enterprise* was returning to Hawaii after a visit to Wake Island, and *Saratoga* was on the West Coast.

In some respects, despite the devastation inflicted along Battleship Row, good fortune smiled on the Americans. Due to rough seas, Admiral William F. "Bull" Halsey's *Enterprise* task force reduced speed on its way back to Oahu. Otherwise, *Enterprise* might have arrived in Pearl in time to receive plenty of attention from Japanese planes.[14]

Q: *What if U.S. carriers had been at Pearl Harbor on December 7?*

Jon Parshall

Ugly. We could have stood to lose one, probably; losing more than one would have made subsequent operations really, really difficult. On the other hand, considering that we put all but two of the battleships sunk at Pearl Harbor back into commission, there's nothing that says that one of our carriers caught at dockside would have been damaged beyond recall, either.

Keith Allen

We certainly could have suffered a severe blow, although of course the extent of damage to *Enterprise* and *Lexington* (the only two carriers operating out of Pearl at the time) is speculative. Had they been lost or knocked out for a long time, as is likely, we probably would not have launched the various carrier raids in early 1942, including the Doolittle strike. We would still have had *Saratoga*, *Yorktown*, and *Hornet*, although *Saratoga* was put out of action by a submarine in January 1942. With only *Yorktown* and *Hornet* available, Nimitz would have been highly reluctant to commit carriers to stop the Port Moresby operation, but would probably have had no choice but to fight at Midway. Operations in the south and southwest Pacific would have focused on Papua, not the Solomons, if the Japanese had gained a foothold at Port Moresby.

Dusty Kleiss

Had several carriers been at Pearl Harbor on December 7, the Japanese would have made numerous scheduled attacks. All of our oil tanks would have been demolished, our repair facilities would have been demolished, and thousands more lives would have been lost, especially if we were in Battle Condition 13 (state of readiness). Oahu would have been out of business for at least a year, and might even have been occupied eventually by the Japanese. Midway would easily have been occupied by the Japanese and used as a submarine base.

Even if just the *Enterprise* had been in port at Pearl Harbor on December 6, as had been originally scheduled, the result could have been devastating for the United States. The Japanese most certainly would have sunk the *Enterprise*. Without the *Enterprise*, and her experienced sailors and aircrews, the U.S. Navy would not have been able to carry out offensive operations against the Japanese during the first half of 1942, as actually happened.

Crossing the Pacific

What if the IJN had not raided Hawaii but instead forced the Americans to carry the war to Japan? The Navy General Staff, Nagumo, and Nagumo's chief of staff, Ryunosuke Kusaka, thought the Pearl Harbor operation was too risky. They preferred to use Japan's carriers against the Philippines and the Dutch East Indies, in hopes of fighting a decisive battle with the U.S. Navy closer to the Japanese Home Islands.[15]

Q: *If Japan had attacked the Philippines but not Hawaii, how would the U.S. Pacific Fleet have responded? Would it have met with disaster in the central Pacific?*

The burning USS Arizona *on December 7. Could the* Arizona *have been lost instead in the central Pacific if the Japanese had not attacked Oahu?*

Stephen D. Regan

The Japanese mistakenly attacked Pearl Harbor because of a poor understanding of the American political situation and the U.S. electorate. Japan perceived that the Americans would fulfill their commitment to their allies, especially to the British, and to their own protectorate, the Philippines. The decision for a brilliant and surprising attack on Pearl Harbor and the elimination of a U.S. naval threat in the Pacific was the only viable option, and one that was played to near perfection.

They were wrong. The fact that the United States lost the Philippines even after the war was commenced provides an abundance of evidence that the Philippine Islands were a lost cause. Obviously, General Douglas MacArthur was ill prepared for an attack or even a solid defense of the islands in spite of his egotistical and grandiose boasts.

The Philippines attack would have started with an aerial attack on important military sites including landing fields, harbors, military installations, and headquarters followed immediately by an invasion by the Japanese army.

MacArthur would have ended up on the run just as happened in the war, except that he would have been defeated earlier and with greater losses.

The United States would have been incredibly aggrieved by the loss of the Philippines, Roosevelt would have economically quarantined the islands of Japan, but he did not have the power to demand war. Senator William Borah and his powerful band of vocal isolationists kept America from going to war to save England, and it is clear that he would not have allowed the United States to enter a war with Japan over some distant islands only tangentially connected with the United States. His "America First" attitude would have also aroused little concern on the loss of Hong Kong, Singapore, and French Indochina. The Republicans were opposed to any semblance of foreign war and certainly would not have been swayed by FDR's Asian policy.

Had the Japanese taken the Philippines and progressed down the South China Sea toward Southeast Asia, they would have met only token resistance from the British, the French, and the Dutch, who were already involved with saving their own hides on the home front.

The joker in the deck of this scenario is Russia. Would Russia have invaded Japan? Would Russia have moved into Mongolia, China, and Korea? Would Russia have moved toward the Asian frontier instead of chasing Hitler back across Europe? Would Russia have split its forces more equally between the Japanese and the European theaters?

I am not a Russian expert and cannot answer these questions with any validity; however, I do believe that the Japanese would have avoided war with America by attacking the Philippines instead of Hawaii. One might even ask if the Philippines were very important to Japan in 1941.

Donald Goldstein

Had the Japanese attacked the Philippines and not Pearl Harbor, then the U.S. Pacific Fleet would have been intact and Kimmel would have brought the fleet out and attacked the Japanese. I believe that's something that should have happened at any rate. When the "war warning" was issued, Kimmel should have had the fleet sail.

Had the Japanese not attacked Pearl Harbor, the Americans might have immediately gone on the offensive. Of course, that might have turned out badly had some of the U.S. ships been sunk. Ships sunk at Pearl Harbor could be raised. Ships sunk in the central Pacific could not be raised.

Jon Parshall

We probably would have fought according to War Plan Orange, which meant a general advance across the Pacific, which would likely have precipitated fleet engagements somewhere out there. Given the superiority of Japanese naval air assets at the beginning of the war, it's reasonable to assume that these early engagements would not have gone well for the United States. At the very least, we would have been at a disadvantage. If that had been true, we would likely have lost numbers of our battleships, but in much deeper water, and likely with heavier casualties to boot.

C. Peter Chen

The Japanese naval leadership was very mindful of the presence of the American battle fleet in Hawaii. Had the attack on Pearl Harbor not been conducted at the opening phase of the Pacific War, Japanese victories at Guam, Wake, the Philippine Islands, and Malaya would have been further guaranteed, since one or more of the attacks would have been further supported by carrier-based aircraft. Actual history provides us with how much Mahanian thinking influenced Japanese thinking. With that in mind, had the battle fleet remained intact, the Japanese would likely have searched for a decisive naval confrontation in early to mid-1942, possibly at the expense of delaying Indian Ocean and Midway operations, in an attempt to wipe out the American fleet in the Pacific. Although at this time the Japanese fielded six fleet carriers with top-notch pilots, which overpowered their American counterparts in the Pacific, the outcome for such a Mahanian confrontation could not easily be ascertained. Factors such as location of battle (closer to American or Japanese land-based air forces?), complexity of attack plan devised by the Japanese (overly complex like Midway or Leyte Gulf?), and even luck would all need to be taken into account.

Invading Oahu

Japan successfully invaded several U.S. possessions in 1941–42, including Guam, which surrendered in a matter of hours. However, it took the Japanese twelve days to overcome Wake Island and nearly five months to conquer Luzon, largest and northernmost island of the Philippines.

In December 1941, Oahu boasted a garrison of 25,000 troops. After blasting Pearl Harbor, one has to wonder if the Japanese could have taken Oahu.

In the end though, due to supply concerns and the fact that troop transports would slow down their carrier task force, the Japanese decided not to attempt an invasion.[16]

Q: *Could the Japanese have successfully invaded Oahu after December 7 or perhaps in the spring of 1942?*

Jon Parshall

First of all, it's important to understand that the Japanese considered such an operation impossible. Staff Officer Tatsukichi Miyo of the Plans Division in the Japanese navy's general headquarters (GHQ) specifically mentioned that the Japanese did not have sufficient assault shipping both to invade Hawaii *and* conduct opening phase operations against targets like Malaya. In other words, assault shipping was a very precious commodity (as the Allies would have attested later in the war as well). This is why the Japanese operations aimed against the Southern Resource Area unfolded in sequential fashion: there was only enough assault to lift one or two divisions at a time. Given that the Malayan operation was vital to securing the approaches to the Southern Resource Area (which is why Japan was going to war in the first place), it was almost inconceivable that the Japanese would have dropped these operations in favor of an amphibious landing against Hawaii at the outset.

This was further compounded by the naval GHQ's aversion to the Pearl Harbor raid itself, which was viewed as incredibly risky at best, even bordering on lunacy. Had Yamamoto demanded that an amphibious operation follow up the carrier raid, I think there is every possibility that he would have been asked to resign. At the very least, it is almost impossible to conceive of a scenario where he would have gotten his way, because if the Hawaiian invasion fleet had been destroyed, Japan would have essentially lost the war on Day 1. If the assault shipping and the assault troops had been destroyed, there could have been no Phase II operations against Borneo and Java. That was simply too large a roll of the dice. As a final note, the army would never have countenanced such an operation in any case. If the navy wanted to throw away its carriers off Hawaii, that was one thing. But I can't conceive of a scenario wherein the army would have gone along with this sort of operation.

Pearl Harbor, late October 1941. Could Japanese ground forces have overcome Oahu's defenses?

Even if the Japanese *had* decided to make such an attempt, an invasion on December 7 itself was impossible. The earliest they could have landed would have been on December 8 or 9. Why is this so? Because of the operational speed of the units involved. Nagumo's task force had reached a point approximately five hundred nautical miles north of Oahu (where American air patrols would start to become a matter of concern) at 10:00 a.m. local time on December 6. From this point, it proceeded southward at high speed (24 knots) to reach its takeoff point some two hundred miles north of Oahu by 4:00 a.m. on December 7. In other words, Nagumo had timed his arrival so that his fleet would be placed at the extreme edge of American air patrols, and then could dash in (with much of his final approach done under cover of darkness), so as to minimize his exposure to American search planes.

Nagumo, of course, was able to conduct his transit with warships, all of which had speeds of 28 knots or better. Most contemporary attack transports of the time, though, would have been constrained to 9–10 knots at best. One has to assume that an invasion force would have been situated so as to reach a similar point five hundred nautical miles away from Hawaii at roughly the same time as Nagumo's carrier force—arriving any earlier would, in effect, put them in front of Nagumo's covering force, thereby exposing them to premature detection and destruction by Hawaiian air and naval power before the main attack could be delivered.

From five hundred miles out, they would have required more than two days to reach Hawaii, and from there conduct an invasion. This means the invasion forces could have reached Hawaii, at the earliest, by about noon on December 8—hardly the most propitious time for conducting an invasion. Japanese amphibious doctrine (such as it was) tended to emphasize night-landing operations conducted against undefended beachheads. Thus, I'm inclined to argue that any Japanese landing operation most likely would not have been conducted before the evening of December 8–9, or even December 9–10. This would have given the Americans a minimum of two days to get ready. Of course, historically the Americans mobilized every soldier on Oahu to repel exactly this sort of attack.

Whether the Americans could have repulsed a Japanese landing is open to debate. The American forces were obviously badly rattled by the air attacks on the islands. On the other hand, they were also numerous, and equipped with plenty of firepower. Whether they would have wilted in the face of a divisional-scale landing by the Japanese is unknowable. My inclination, though, is to say they would have held, or at the very least inflicted substantial casualties on the Japanese in the process. As was demonstrated later in the war at places like Alligator Creek on Guadalcanal, Japanese frontal rushes against American firepower most often resulted in catastrophic Japanese losses for little in the way of gain.

The early-war invasion of Wake Island by the Japanese offers some interesting analogues as well. Wake was defended by around 450 military personnel, mostly U.S. Marines. During the final assault, the Japanese deployed a force of 1,500 Special Naval Landing Forces (SNLF) for a nighttime landing. While the attack was eventually successful in terms of capturing the island and forcing an American surrender, during this ground action the U.S. defenders exacted a minimum of four hundred Japanese killed in action, for the loss of about fifty Americans. The Japanese landing force also suffered a very high number of wounded in action. In other words, an American defense force outnumbered 3-to-1 had inflicted casualties on the enemy at a rate of 8-to-1.

Indeed, the official Marine history of the battle makes it clear that had the defenders possessed a single company of Marine infantry and a platoon of light tanks in reserve for a counterattack, the Japanese landings would likely have been annihilated. When viewed in the context of a Japanese invasion of the Hawaiian Islands, where the *attackers* would have been outnumbered, it is difficult to escape the conclusion that Japanese prospects for capturing Oahu, even a few days after a successful air attack against Pearl Harbor, were not good.

The second scenario, an invasion of Hawaii conducted later in 1942, is even more clear-cut, and I will simply quote at length from *Shattered Sword* here, because I've answered this question in depth. There is almost no conceivable set of circumstances under which the Japanese could have been successful against Oahu after mid-December 1941. Once the element of surprise was gone, the Japanese would have lost their most important advantage. And the Americans moved very quickly to reinforce their most important naval base so as to prevent just such a Japanese attack.

By April 1942, the Americans had 62,700 Army troops (two full infantry divisions, plus support troops) in Hawaii, and another 8,900 air personnel.[17] The U.S. Army expected this total to reach at least 115,000 ground and air personnel in the near future. This figure does not include the tens of thousands of U.S. Navy personnel located at Pearl Harbor. Thus, even had the Japanese followed up a victory at Midway in short order with an attack on Hawaii, they would have had to contend with a Hawaiian garrison of at least 100,000–150,000 U.S. servicemen. These American troops were primarily located on Oahu, which was small enough to be defended in depth but big enough to maneuver on, making it an enormously difficult nut to crack.

The Japanese themselves thought that capturing such a stronghold would require at least three of their smaller infantry divisions, or roughly 45,000 troops.[18] This was certainly a conservative estimate, but even so it represented a force ten times larger than the one the Japanese had planned to employ at Midway, and three times larger than they had ever amphibiously lifted at one time. It is doubtful that Japan had the sealift capacity to contemplate such an undertaking across nearly four thousand miles of open water in any case. Nor is it likely that, even if the Japanese were so lucky as to have captured the islands, that they would then have been able to keep their troops in supply, let alone the civilian population of the islands.[19]

Even if sufficient transport could have been found, a Japanese assault force would have to be landed in the face of withering American fire, without much in the way of specialized equipment, and without an effective naval gunfire or air support doctrine. If the Japanese could get ashore at all (a dubious proposition, given the size of Oahu and the depth of the American defenses), there could be none of the bold flanking movements through the jungle that the Japanese had used to such great effect in Malaya. The American defenses would be too well anchored, and the island too densely defended for such gimmickry to work. Hawaii would have to be taken frontally. Given the heavy losses the Japanese had suffered at Wake, mentioned above, it is difficult to escape the conclusion that a Japanese landing on Oahu would have resulted in a bloodbath worthy of the Somme. Granted, the Japanese were hardly squeamish over taking such losses, but it's difficult to see how such an invasion could have been successful.

Furthermore, the Japanese would have to secure air superiority over Hawaii with carrier assets alone. Even had they captured Midway as a prelude to invasion, their land-based aircraft would have been over a thousand miles away, and could have played no real role in such an invasion. Yet unlike the carrier task forces the Americans would go on to employ in 1944, even at the height of its powers Kido Butai never had the ability to stand off an enemy's island bastion for weeks on end and beat it into submission. In the first place, Kido Butai couldn't bring a sufficient number of aircraft to get the job done. By April 1942, Hawaii boasted 275 combat aircraft, a figure which had increased as the battle at Midway had loomed. In the event of an American defeat there, Hawaii's air force could have been augmented still further by naval aircraft shuttled in from *Saratoga* or *Wasp*, much as the Americans went on to do at Guadalcanal. This meant that Kido Butai, even with all six carriers available, would have fought against Hawaii from a position of numerical parity at best. But more important, the logistics for a sustained Japanese carrier presence off the shores of Hawaii simply weren't there. Kido Butai could mount raids, but it could not project sustained power ashore. This meant that Japanese ground forces, even if they managed to stay ashore, would probably lack consistent air cover while trying to conduct offensive operations—not a good recipe for ultimate success.

It is true that if the Hawaiian Islands couldn't be captured, the Japanese might have tried blockading them instead, using a combination of submarines and surface forces. But Japanese doctrine said that submarines weren't supposed to be used to hunt down merchant shipping, and as a result their subs never proved as effective in this role as they might have been. For their part, Japanese surface forces couldn't hope to operate in the face of American land-based airpower without Kido Butai to support them. Yet the presence of Japanese carriers in the area would have necessarily been sporadic. Even in the face of a concerted blockade, it is almost impossible to imagine the Americans being willing to sacrifice the garrison there, not to mention the large civilian population. If maintaining the logistical flow to the islands necessitated sustaining Murmansk-like losses in the supply convoys, so be it, but the convoys would still have been sent. America was engaged in a total war, and failure was not an option. Just as the British had risen to the London Blitz and defended Malta in the Mediterranean, and the Russians had endured the seemingly endless siege of Leningrad, so too would the Americans have been determined to hang on to Oahu—their last outpost in the Pacific. The only logical conclusion one can reach from all this is that Hawaii was largely impregnable. Its value as an American naval base might have been diminished in the short term, but the islands themselves could never be taken.

Frank Shirer

This question is the classic one that the intelligence analyst gets into trouble with his commander over: What is the enemy's capability versus its intent (goal)? Yes, the Japanese could have invaded Hawaii, if the U.S. fleet had been rendered ineffective, but why would they have wanted to do so?

Answering this question involves the consideration of what was the primary goal of the December 7 attack, how did it fit into overall Japanese strategy, and was an invasion not only logistically possible but was an occupation force logistically sustainable over a period of several years? On all points, I consider the answer to be, "No." Pearl Harbor was the temporary home of the U.S. Pacific Fleet after it had been ordered to deploy from its West Coast homeports to the forward base of Pearl Harbor in 1940. The move was part of President Roosevelt's diplomatic pressure on Japan to end its war with China and withdraw from all conquered territory. The leadership of the Japanese navy saw this move as a threat, to oppose militarily any Japanese moves to gain the oil and other resources of the East Indies. Admiral James O. Richardson, Admiral Husband Kimmel's predecessor, as commander in chief, Pacific

Fleet, was relieved of command in January 1941, because he opposed leaving the Pacific Fleet in Hawaii. He objected because repeated U.S. Navy war games in the 1930s had demonstrated the fleet's vulnerability to attack by an enemy carrier force.

The decision was made to make a strike with the majority of Japan's aircraft carriers and inflict maximum damage on the American fleet. After the attack, the carriers were needed to provide air support for the Japanese invasion forces assigned to conquer Malaya, the Dutch East Indies, and the Philippine Islands. A decision to invade and seize the Hawaiian Islands would have required a much larger fleet to sail from Japan, one which included troop transports and cargo ships, ships that were also needed to carry troops and supplies to support the other invasions that were being planned. The Japanese navy was stretched to its maximum in accomplishing the coordinated attacks that actually occurred. Committing the troops and transport for an invasion of Hawaii would have caused shortages elsewhere and possibly ensured the failure of another attack.

Also, what would have been gained by invading Hawaii in December or at another time? Japan needed oil and other raw materials for its war machine; Hawaii had none of these. Agriculturally, Hawaii was and continues to be an importer of food for its civilian population. The Japanese would have had to send food to Hawaii for the civilian population and for their own occupation forces. Even turning the American population into a slave-labor force to raise food would not have worked because there was not enough cultivable land; thus the islands would have been a drain on Japan's scarce resources. William Forstchen and Newt Gingrich's two-volume alternate history, about the attack on Pearl Harbor and a subsequent invasion and occupation, provides a realistic example of this. The denial of Hawaii as a forward operating base for the U.S. fleet would have been better performed by use of a submarine cordon around the islands, attacking fleet and cargo vessels bringing troops, war supplies, and food to the islands.

An invasion of Hawaii did not provide either an immediate or a long-range advantage. The invasion force that was part of the Midway operation was to occupy that island, and this force would have been easier to support than the Hawaiian Islands. The goal of capturing Midway, after sinking what remained of the U.S. Pacific fleet, was to have a location from which long-range surveillance and harassing attacks could have been made on Pearl Harbor. Also, it would have been a "threat" that would have diverted some attention away from Japan's drive toward Australia. The long-term occupation of the Hawaiian Islands was not a logistically supportable option.

Clayton K. S. Chun

Japanese forces would have had a difficult challenge to take a fortified Hawaii just after the Pearl Harbor raid. Although it was not impossible, Tokyo faced significant obstacles to launch, win, and sustain an amphibious invasion and occupation of Oahu. A successful invasion of Hawaii would have provided a huge strategic advantage to Japan. Such an event would have denied the American military a vital logistics base to support operations in the Pacific for an extended period. Japanese forces could have swiftly isolated American and Allied forces throughout the Pacific. Additionally, Washington would have required a tremendous effort to retake a Japanese-controlled Hawaii from the West Coast, hundreds of miles away.

There were many problems associated with invading Hawaii. First, did the Imperial Japanese Army and Navy have the sufficient and appropriate amphibious capability to land forces to defeat defensive forces in Hawaii and occupy the territory? Japanese military forces did invade Malaya, the Philippines, Guam, Thailand, and Hong Kong immediately after the Pearl Harbor raid. However, many of the landing forces were regimental in size and originated from French Indochina, Formosa, and China, close to their targets. At the time, the Imperial Japanese Navy did not have to support a lengthy amphibious transit across the Pacific.

Second, a follow-on land invasion of Oahu would have required the defeat of the Hawaiian Army Command forces, which included two infantry divisions and extensive coastal defenses, a weakened Hawaiian Air Forces, and a Pacific Fleet that still contained aircraft carriers and a still-potent surface force to defend the islands. These defenses would have required a significant offensive force to overcome.

Third, given the Japanese force needed to initiate and launch an attack, what would Tokyo have had to give up? Japan did not have sufficient amphibious capability to conduct simultaneous amphibious operations in Southeast Asia, Hong Kong, Hawaii, and other areas. If Tokyo had focused on a Hawaiian invasion, then the delay of an incursion into Southeast Asia and other areas would have been costly. Japan still required oil and raw materials to arm and equip its forces in China, keep an economy running, and support the invasion. British and American forces in Malaya and the Philippines could interdict Japanese naval and merchant marine movements in the region. Japan would have risked crippling its ability to continue the war. An early Hawaiian invasion would have delayed the acquisition of resources, and that could have affected all theaters of war for the Japanese.

Fourth, could the Japanese have sustained an extended campaign to take Hawaii? Tokyo's limited merchant marine fleet would have been strapped to support its forces so far from its Home Islands. Japanese forces did not immediately control support bases, like Wake or Midway Islands. Taking time to take those islands would have allowed the United States to reinforce Hawaii with additional air, land, and naval forces. A long, vulnerable Japanese logistical tail would have become a lucrative target to American submarines and other forces. Similarly, the Japanese would also have required extensive military forces and logistics to administer an occupied Hawaii. For the Japanese, the distance from their Home Islands to Hawaii is much greater than the distance from the U.S. West Coast to Hawaii. In some respects, the Japanese forces that would have stayed in Hawaii would have been like those who remained in the Aleutians after Midway.

Jim Hallas

The success of the Japanese surprise attack on Pearl Harbor on December 7, 1941, has come to be regarded as inevitable. In reality, the attack was bold to the point of recklessness and could easily have failed.

It must also be remembered that the attack was not remotely on the scale of the amphibious assaults mounted by the United States later in the war. Approved only with great reluctance by the Imperial Navy's General Staff, the Pearl Harbor attack was a raid, pure and simple, a quick in-and-out strike with success depending on surprise and speed. Slowing the strike force down with the addition of plodding supply and troop ships—even if those additional ships and troops had been available in light of other military commitments at that time (which they were not)—would have incurred an unacceptable risk of discovery and destruction. As a result, a ground invasion of the Hawaiian Islands on December 7 was simply not feasible.

The potential for a successful invasion months later, even granting the seizure of Midway by Japanese forces, was also remote. The Japanese estimated it would take three infantry divisions—about 60,000 men, give or take—to invade Hawaii. Considering that by early 1942 the United States had two infantry divisions on the islands, as well as tens of thousands of sailors, Marines, and support troops—perhaps as many as 100,000 trigger-pullers all told—that estimate seems optimistic to the point of foolishness. In any case, the Japanese could not spare three divisions.

Secondly, a successful invasion would have necessitated control of the air and surrounding sea, a difficult proposition at best with U.S. forces on alert and operating from unsinkable land bases.

And finally, the Japanese lacked sufficient cargo ships to launch and sustain an invasion of that magnitude.

Still, the implications of such an invasion are interesting to contemplate. At the least, it could have seriously impacted the United States' "Germany First" strategy. It also would have offered the potential for large-scale fleet actions with Japanese naval forces early in the war. Hindsight indicates the United States was not particularly well prepared for such confrontations and could well have come out on the losing end. Such actions could have dramatically altered the course of the war, though not the eventual outcome.

Nimitz

Ten days after Pearl Harbor, heads rolled. Vice Admiral William S. Pye replaced Kimmel as commander in chief of the U.S. Pacific Fleet.

More changes were coming. On December 30, Admiral Ernest J. King took over as commander in chief, U.S. Fleet; the following day, Admiral Chester W. Nimitz became the permanent CinC, Pacific Fleet.

King directed Nimitz to focus on two priorities: holding the Midway-Johnston-Hawaii line and maintaining the line of communications between America and Australia.[20]

Q: *What if Nimitz had been CinC, Pacific Fleet before December 7? What if someone else had been appointed CinC, Pacific Fleet after December 7?*

Jon Parshall

Nimitz most likely would have been taken by surprise, just like Kimmel was, in which case he would most likely have been scapegoated, just like Kimmel was, and the United States would have lost an exceedingly good admiral.

It needs to be understood that the failure at Pearl Harbor on the Americans' part was less one of poor planning than of poor imagination. No one in the U.S. Navy had any idea that the Japanese navy was capable of conducting multi-carrier operations, for the very simple reason that the U.S. Navy itself had not yet made the conceptual leap toward these sorts of carrier operations. It's pretty tough to plan against contingencies that you don't even remotely understand. The most that someone like Kimmel or Nimitz might have planned against would have been a raid by one or two carriers, which most likely could not have inflicted more than nagging damage to a complex as large and sophisticated as Oahu. This was, after all, how we operated our own carriers—in penny packets. Given the shallowness of the harbor, even an operation like the one the British conducted against Taranto would most likely have been viewed as nigh on impossible.

Bear in mind, too, that there was no intelligence on what the Japanese were capable of in this regard. A lot of this has to do with the very rapidly evolving state of Japanese carrier doctrine in late 1941. As discussed previously, the Japanese had to develop multi-carrier doctrine very rapidly, as First Air Fleet went through its workups aimed at the Pearl Harbor operation. It's asking a lot of U.S. naval intelligence to be able to detect these doctrinal developments in time, let alone understand their operational significance, and then get that understanding out to the fleet. This is doubly true when your Navy has yet to make that same conceptual leap.

Now, it may be that Nimitz (who was an excellent officer) might have raised the overall level of alertness on the island. But I think it's asking a lot for Nimitz to be able to predict, and then guard against, the size and sophistication of attack that the Japanese were going to bring to the islands. The Japanese had developed a revolutionary weapons system. Enemies on the receiving end of such a system are rarely successful in defending themselves against it during its debut. With all due respect to Chester Nimitz, whom I consider the finest admiral of the war, I rather doubt that he would have been able save Pearl Harbor from destruction.

Stephen D. Regan

The U.S. Navy was very seniority oriented, as we saw at Midway when Frank Jack Fletcher was named to tactical command of the two task forces involved. Kimmel was in the senior position for CINCPAC (commander in chief, Pacific Fleet) or similar duties. Wilson Brown or even William Halsey would have rated the command before Nimitz; therefore, the question of his being at Pearl Harbor on December 7, 1941, is a moot point. Even when he was elevated to that post, it was a leapfrog move over several more senior officers.

The debate about Pearl Harbor readiness under Husband Kimmel's command versus the possible readiness under Nimitz is interesting, but the end result probably would have remained the same. The attack on Pearl Harbor was simply beyond the visceral comprehension of virtually everyone in the Navy, even though most officials had a cognitive recognition of Hawaii's vulnerability. No doubt, Nimitz or anyone else would have been as much in the dark as Kimmel. Nimitz would have been caught with his pants down just as Kimmel was.

Everyone from the president of the United States to the greenest recruit knew that war with Japan was imminent, but the exact date and nature of the initial attacks were debated in every bar, coffee shop, and military/governmental meeting from Honolulu to Washington, D.C. A commander cannot keep his ships on constant alert and a captain cannot keep his crew at battle station seven days a week. Nimitz would have suffered the same fate as Kimmel had he been in charge.

Fortunately, the United States was placed in a position whereby standard operating procedures were easily ignored. Virtually no one else other than Nimitz was as good at balancing the politics, the military readiness, and the ambition to lead America in a Pacific war. We were extremely lucky that Chester Nimitz was available for CINCPAC after the attack.

The discussion of other admirals being CINCPAC instead of Nimitz is viable because he certainly was not the senior man for the position, and his selection was something of a surprise to everyone. His selection offers evidence that Ernest J. King and his predecessor Harold "Betty" Stark knew exactly who was the best man for the job. King needed someone with a more pleasant temperament to counter his own acerbic nature yet someone with vision, courage, and intelligence to run that theater. In another century, Chester Nimitz would have been canonized.

It is hard to imagine someone else as CINCPAC than Nimitz. Many of the admirals available for command did not come off well in the early months of the war. Raymond Spruance was too inexperienced and too far down on the seniority list, W. S. Pye was too closely linked to Kimmel, Frank Jack Fletcher was both unlucky and unsupported by the aviation community, Wilson Brown was more of an administrative type than a leader, Harold Stark was tainted by the Pearl Harbor disaster, Robert Ghormley lacked offensive thinking, William Halsey exhibited initial mental weaknesses shown by his constant nervous outbreak of hives, and everyone else was too junior for the job. To mix metaphors, combat eventually weeded out the weak links and illuminated the rising stars. By the end of 1942, most of the right people were in the right positions.

Summary

Japan had gambled and won. By smashing the U.S. Pacific Fleet at Pearl Harbor, the Japanese were now in position to finish off the Philippines and the Dutch East Indies without having to worry about American interference.

However, Pearl Harbor's ship-repair facilities and oil stores were left untouched, and America's aircraft carriers had escaped destruction. By attacking a few days earlier or later, the Japanese might have caught one or two U.S. carriers in port. Without *Enterprise* or *Lexington*, the Americans would have had a difficult time contesting Japanese operations in the spring of 1942.

December 7 had given the Japanese a free hand and they quickly took advantage, acquiring land and resources. Holding on to the new possessions would be another matter.

Four

RISING SUN: DECEMBER 1941–APRIL 1942

WITH THE AMERICANS REELING, the Imperial Japanese Navy sailed relatively unopposed in December 1941, conquering British, Dutch, and U.S. holdings at an alarming rate. Guam surrendered on December 10. After heroic resistance, Wake Island fell on December 23, leaving the Philippines as the only U.S. possession in the western Pacific not in Japanese hands by the end of 1941.

Saving Wake

Just hours after the Pearl Harbor attack, Japanese bombers put in their first appearance over Wake Island. Four days later, the IJN showed up with a fleet that included three light cruisers, six destroyers, and two transports. Wake's defenders, including 447 Marines under the command of Major James Devereux, were not about to go down without a fight. Between artillery and bombing runs made by Wildcat fighters, the Marines sank or damaged at least two light cruisers, four destroyers, and a troop transport.[1]

Less than two weeks later, on December 23, the IJN returned in even greater numbers, with six heavy cruisers, carriers *Soryu* and *Hiryu*, and nearly two thousand elite Marines. After several hours of hard fighting, Wake's garrison capitulated.[2]

The naval brass at Pearl Harbor had hoped to reinforce Wake with a relief force screened by *Saratoga* and commanded by Rear Admiral Frank Jack Fletcher, while task forces built around *Enterprise* and *Lexington* provided support.

However, the plan unraveled when Fletcher ran into delays due to refueling and zigzagging to avoid subs. Orders for *Saratoga*'s planes to search for enemy ships and for the seaplane tender *Tangier* to evacuate Wake's garrison wound up being canceled. Back at Pearl, Vice Admiral William Pye had received a dispatch from Admiral Harold Stark, chief of naval operations, in which Stark labeled Wake Island as a liability.[3] That settled matters. Pye ordered *Saratoga* and *Lexington* back to Hawaii, sealing Wake's fate.[4]

Q: *What if the Wake Island relief effort had not been canceled?*

Jim Hallas

The Wake Island relief mission, as it developed, lacked definite direction and efficient coordination. It is little wonder that it ended up as a frustrating—and futile—round-trip to nowhere.

To have achieved any element of success, the effort would have required a more decisive approach from the start. First, a realistic goal or combination of goals should have been established. Was the purpose to: 1) reinforce Wake, 2) relieve the beleaguered garrison, or 3) strike directly at Japanese naval forces?

Secondly, Admiral Husband Kimmel, who initially launched the mission before his relief by Admiral William S. Pye, should have acted decisively and committed sufficient force to the effort. Instead of sending two or all three of the available carriers toward Wake, Kimmel committed only *Saratoga*. The *Lexington* was sent off on a useless diversion toward the Marshalls and *Enterprise* hung back as an ineffective reserve.

Finally, to have had any chance of success, the relief force should have been directed to make all possible speed to Wake. Instead, Task Force 14 took a leisurely pace in order to allow its slow oiler to keep up and wasted valuable time conducting unnecessary refueling. The task force was still well off the island when the Japanese landing took place. Fearing a trap, Admiral Pye scrubbed the mission.

Pye has been roundly criticized for halting the relief effort, but his caution at this stage of events was probably wise. He knew little of enemy dispositions beyond the fact that Japanese carriers were on the scene. Due to the slow approach and dispersion of U.S. carriers, only *Saratoga* was even marginally within striking range of the enemy naval force. U.S. forces would have been committed piecemeal—and if the *Saratoga* had run into trouble, she would have been on her own.

It is also questionable whether *Saratoga* was a match for Japanese forces in the area. Given surprise, her aircraft presumably could have inflicted damage on enemy shipping congregated around Wake. But then she would have had to deal with two Japanese carriers, *Soryu* and *Hiryu*. Her fighter squadron was understrength with only thirteen Wildcats. She also carried fourteen Marine Corps Brewster Buffalo fighters. The latter, intended for the defense of Wake, were of doubtful value. Judging from later events at Midway, the obsolete Buffaloes would have been slaughtered by Japanese fighter pilots.

Saratoga did not have sufficient antiaircraft batteries, and her antiaircraft ammunition was not reliable. And finally, her pilots would be going up against an experienced enemy—arguably the best naval aviators in the world at that time. Instead of inflicting a defeat on Japanese carriers and saving Wake, *Saratoga* might well have ended up at the bottom of the Pacific.

If Kimmel and Pye wanted a fight, they should have committed sufficient force in the first place. However, there is still no guarantee that even all three U.S. carriers would have prevailed. The situation was confused. This was no Midway where the Japanese plan was known in advance. While naval officers on the *Saratoga* were so eager to close with the enemy that Admiral Aubrey Fitch had a near-mutiny on his hands when the recall order came, the Japanese were a competent and dangerous enemy. Many hard lessons lay ahead for the U.S. Navy; it would have been disastrous to have suffered a major defeat at Wake on top of the debacle at Pearl Harbor.

Plans to reinforce Wake were also unrealistic. Considering events at Pearl Harbor and the proximity of Japanese forces in the Marshall Islands, Wake had become untenable. A better option would have been to rush an evacuation force to Wake in the window between the first Japanese failed assault on December 11 and the second landing on December 23. Safely retrieving the garrison after its heroic defense of the island would have provided a major morale boost, avoided more damage to the fleet, and sacrificed nothing of real strategic value. Once that window closed at Wake, the wisest course, hard as it may have been, was to leave the garrison to its fate. As an old military axiom warns, "Never reinforce defeat."

Stephen D. Regan

The fog of war permeated the Pacific as CINCPAC Husband Kimmel was relieved by William Pye who was to be relieved by Chester Nimitz, and Harold "Betty" Stark was being relieved by Ernest King as CINCUS (pronounced "SINK-us" but rapidly changed to "COMINCH"). Pye was aware of the fact that he had only two oilers and three carriers, and he was greatly worried about becoming involved in a major battle with such limited resources when he would not be the person who would have to cope with the possible defeat of the task force.

Contrary to popular belief, Frank Jack Fletcher was very much aware of the problems facing the Navy in the Pacific. He was also aware that the *Saratoga* and the oiler were, bluntly stated, more important than Wake Island. He was very careful to keep his ships fully fueled in case of the loss of the oiler, and he was concerned about the first major naval battle since World War I with the United States on the weaker side. Nevertheless, Fletcher was ready, willing, and able to proceed to Wake Island. The recall of the rescue task force was not his doing nor was it in agreement with his desires in spite of the fact that he was held responsible for this failure.

It must be remembered that Admiral King had no love for Fletcher and he did indeed believe Fletcher to be culpable. King never let facts interfere with his biases. Pye ordered the return of the task force.

If Fletcher had indeed proceeded to Wake, it is incredibly possible and downright probable that the early war would have been significantly different. Fletcher would have entered the battle only by skimping on fuel for his other ships. The Japanese were no fools, and they would have attacked the carrier and oiler with a vengeance. *Saratoga* had little with which to protect herself, and her aviators had the twin missions of keeping the Japanese away from Wake Island while also covering the ship.

In war game exercises, it was fairly apparent that should a war with Japan commence, Wake Island was of minor significance. It had commercial value for Pan American Airways but little value other than that. Pye, Kimmel, Nimitz, Fletcher, and even King himself knew that rescuing the men on Wake was hardly worth the risk. Had Fletcher succeeded in reaching Wake he would have lost the battle anyway. Damned if he did; damned if he didn't.

The business of war is cruel. Decisions that resulted in death, injury, and capture were made daily. Recalling the rescue task force was correct but not heroic. The Marines never forgot Wake nor forgave Fletcher. They would blame much of the loss on Guadalcanal on him as well. The Marine Corps' official history castigates Fletcher. They are wrong. But who can blame them? However, the Marine Corps commandant himself would have made the same decisions.

Unprepared in the Philippines

Shortly after the Kido Butai smashed U.S. battleships at Pearl Harbor, Japanese bombers from Formosa (now Taiwan) flew over Clark Field on Luzon, an American air base northwest of Manila. Surely, Pearl Harbor had left Americans throughout the Pacific on full alert, including forces led by Lieutenant General Douglas MacArthur, commander of U.S. Army Forces in the Far East, based on Luzon. However, when the Japanese arrived over Clark, they were astonished to find enemy planes parked neatly along the runway. The Japanese had a field day, damaging or destroying a large number of B-17 bombers and P-40 fighters, and all but eliminating Clark Field as a B-17 base.[5]

Miscommunication and hesitation hampered the Americans. Upon learning of the raid on Oahu, the commander of MacArthur's air force, Major General Lewis Brereton, sought permission to launch B-17s against Formosa. Unable to meet with MacArthur, Brereton approached MacArthur's chief of staff, Brigadier General Richard Sutherland, and learned that MacArthur had yet to approve offensive operations. For some reason, MacArthur delayed making an immediate attack.[6] By the time he telephoned Brereton hours later and authorized a bombing mission, the Americans had lost the initiative.[7]

Japanese troops landed on Luzon on December 22 and in less than two weeks had Manila. On April 9, with their ammunition and antimalaria medicine in short supply, American and Filipino defenders on the Bataan Peninsula surrendered. Japan completed the conquest of Luzon by taking the island fortress of Corregidor on May 6.[8]

Q: *What if Brereton and MacArthur had communicated more effectively on December 8?*

Bill Bartsch

When news of the Pearl Harbor attack reached Brereton at about 4:00 a.m. on December 8, the Far East Air Force commander, expecting an attack by the Japanese from Formosa anytime after daylight, headed to MacArthur's headquarters immediately. But MacArthur's chief of staff, Sutherland, told Brereton that MacArthur could not see him—he was in conference, Sutherland maintained. Brereton was thus unable to get MacArthur's approval for the strike of his B-17s at Clark Field on Formosa, which he wanted to mount right away. At 7:15 a.m., on a return to the United States Army Forces in the Far East (USAFFE) headquarters, Brereton was again rebuffed by Sutherland, who informed him that MacArthur wanted the Japanese to make the first overt act—he apparently didn't regard Pearl Harbor as qualifying. Only at 10:14 a.m., after he knew of the attacks on northern Luzon by Japanese bombers, did MacArthur give Brereton the go-ahead for a Formosa bombing mission. In the meantime, a massive strike force of 195 Japanese navy bombers and Zero fighters was on its way south from Formosa, its targets Clark and Iba fields.

If MacArthur had received Brereton during his 5:00 a.m. visit and approved his Far East Air Force (FEAF) commander's request—which was in line with the instructions MacArthur received by cable from Chief of Staff George C. Marshall at 5:30 a.m. to put the Rainbow 5 War Plan into effect—Brereton could have had his seventeen in-service B-17s at Clark Field fully armed and ready for takeoff by 6:30 a.m. for an attack on Takao field and harbor. They would have reached Formosa before the 9:30 a.m. takeoff of the eighty-one Betty and Nell bombers and fifty-three Zeros at Takao, delayed on their attack mission by fog that didn't begin lifting until 7:50 a.m.

The Japanese 11th Air Fleet headquarters were expecting an attack on their Formosan airfields at Tainan and Takao by the B-17s anytime from 7:00 a.m. on the assumption that MacArthur had been informed of the Pearl Harbor attack, but they were helpless to get their strike force off the ground due to the fog. With only twenty-four obsolete Type 96 fighters left to defend the fields and an inadequate antiaircraft defense, they feared a B-17 attack on their fuel- and bomb-laden aircraft would not only force them to abort the mission but deal a disastrous blow to their fighter and bomber force.

Even discounting any damage the Clark-based B-17s could have inflicted in an attack on Takao field, the bomber crews—at the time about to leave on their Formosa mission—would have at least been airborne at the time the Japanese struck Clark Field at 12:30 p.m., and thus would have escaped their fate to be caught on the ground, along with the twenty-three P-40s also based there. Indeed, any takeoff before 12:15 p.m. would have saved them.

However, even if Brereton's B-17s at Clark Field had not been decimated in the Japanese attack, they—and the other sixteen B-17s on Mindanao to the south—would not have affected the outcome of the Philippines campaign in any meaningful way. There were just too few of them in the first place, intended only as the first installment of 272 heavy bombers that the War Department was planning to send to the Philippines. Unfortunately for MacArthur, the Japanese didn't wait for this buildup to be completed by April 1942, sticking to their own timetable for southern operations to begin on December 8, 1941, with a strike to wipe out his air power on Luzon right from the start.

John Burton

One can readily appreciate how a change in events triggered by the deviation of an hour, or several minutes, or even a few heartbeats and the blink of an eye might alter history forever. Many students of the Pacific War have postulated that the inexcusable lack of communication between the USAFFE commander in the Philippines, Lieutenant General Douglas MacArthur, and his air force leader, Major General Lewis Brereton, led directly to the disastrous loss of much of Brereton's Far East Air Force on December 8, 1941. Some have gone so far as to declare the event to be "MacArthur's Pearl Harbor."

Regardless of any effect on the outcome of that day, it is difficult to reconcile MacArthur's apparent refusal to meet with Brereton—or the possibility that MacArthur's chief of staff, General Richard Sutherland, neglected to deliver appropriate messages to the USAFFE commander—with anything other than a failure in command. However, was this potential "dereliction of duty" the fatal flaw in the chain of events that led to FEAF's loss of twelve B-17 Flying Fortresses, three dozen first-line fighter planes, and more than a dozen other military aircraft during the Japanese bombing raid, which occurred at 12:30 p.m. on December 8? Could that result have been changed by more prompt or intimate dialog between MacArthur and Brereton? Would any difference in events on December 8 have affected the course of the battle to defend the Philippines from a Japanese invasion?

One can safely assume that neither MacArthur nor Brereton had access to any information that could be construed to provide advance knowledge of Japanese intentions regarding the Philippines. Therefore, when each was awakened shortly before 4:00 a.m. with confirmed news about the attack on Pearl

Harbor (which began at 2:28 a.m. Manila time), both men were starting with a "clean slate." General Brereton was promptly summoned from his quarters at Fort McKinley by General Sutherland, and instructed to attend a meeting at 5:00 a.m. in MacArthur's office inside Fort Santiago.

Before leaving for Fort Santiago (about a fifteen- to twenty-minute drive from his quarters), Brereton made telephone calls from his residence to FEAF headquarters (HQ) at Nielson Field (adjacent to Fort McKinley) and to Lieutenant Colonel Eugene Eubank at Clark Field. Eubank was told to fly down to Nielson Field to meet with the FEAF staff for a war conference at 6:00 a.m. At FEAF HQ, the Interceptor Command, under the leadership of Colonel Harold "Fighter" George, was ordered to place all fighter squadrons of the 24th Pursuit Group onto "full-alert" status, with pilots manning their planes. From this point, the Interceptor Command was intended to be self-sufficient; no further action was expected of General Brereton regarding the fighter defenses. With input from the information center at Nielson, the 24th Pursuit Group commanding officer, Major Orrin Grover, would control his squadron assignments from his post at Clark Field using radio signals.

Lieutenant Colonel Eubank ordered the men of his 19th Bomb Group into action, preparing their B-17s with bombs and ammunition for the trip to Takao. He promptly canceled two of three "visual reconnaissance" flights previously scheduled to patrol the Strait of Luzon. The remaining B-17, piloted by Lieutenant Hewitt Wheless, was to be sent north to Formosa at dawn, for a "weather" reconnaissance. With two planes in the hangar for service and painting, and one more plane completely out of commission (awaiting shipment of parts), Eubank could expect to put fifteen Flying Fortresses (from the 28th, 30th, and 19th Group HQ Squadrons) in the air for a bombing mission.

In the meantime, Brereton would travel to Fort Santiago to meet with MacArthur and secure his permission to execute a pre-planned, retaliatory B-17 strike at Japan's large port facility at Takao (Kao-hsiung) Harbor—an expected departure point for any invasion fleet with intentions on a Philippine destination. This was the *only* mission for which Eubank's bombers could be reasonably prepared without first conducting a preliminary photo reconnaissance over Formosa. Target intelligence was not available for any other location in Japanese territory—apart from the Home Islands of Japan.

When General Brereton arrived at Fort Santiago, he was met by General Sutherland, and told that MacArthur was in conference "behind closed doors."

Brereton discussed his intentions with Sutherland, who agreed that the raid on Takao was sound in principle. However, Sutherland would not sanction the plan until General MacArthur personally reviewed it. MacArthur, in his prewar discussions with Philippine president Manuel Quezon, had tacitly agreed not to launch tactical attacks against Japanese territories from Philippine soil unless Philippine possessions had already been attacked by Japanese forces. Although MacArthur represented the U.S. Army, he was also, in essence, a highly compensated "employee" of the Philippine government—working for Quezon. While Pearl Harbor had been attacked three hours before, the Philippines had not yet been attacked. The available record seems to indicate that MacArthur still had some doubt about the "war status" of the Philippines at 6:00 a.m.

Up to that point on the morning of December 8, direct dialog between Brereton and MacArthur would have made no impact on events. A short while after 6:00 a.m., the wheels of fate began to roll inexorably out of line for the Americans and Filipinos. Delays and confusion resulting from a lack of timely conversation between the two generals had a cascading effect, which arguably resulted in the immediate loss of the dozen B-17s (and damage to several others) during the Japanese bombardment of Clark Field.

Discussions at the Brereton and MacArthur level should have had no bearing at any time on the deployment of fighter aircraft—or the possible outcome of any interception activity; so, one must realize the only history that might have been changed was that of the 19th Bomb Group.

Let us suppose that whatever activity General MacArthur was engaged in between 5:00 a.m. and 6:00 a.m.—which was of such great importance or high sensitivity that General Brereton could not have been included—lapsed momentarily. At that small break, Brereton could have easily explained his plan to MacArthur and probably secured his verbal approval in less than fifteen minutes. There are, of course, a number of other times after 6:00 a.m. that morning when such a discussion might have been taken place prior to 10:14 a.m., when Brereton and MacArthur actually did talk via telephone.

Brereton would have been able to return to Nielson Field for his conference just after 6:30 a.m.—about the same time that Eubank and his executive officer, Major Birrell "Mike" Walsh, arrived from Clark Field. After a brief status meeting, mission orders would have been issued. Eubank and Walsh would have flown back to Clark to conduct a final briefing for aircraft commanders and navigators. The departure of B-17s from Clark could probably have commenced by about 7:30 a.m. With some luck, the American bombers would have reached the target area on the island of Formosa between 10:00 and 10:30 a.m. If Brereton's staff had decided to launch a concurrent photo reconnaissance operation, one of the bombers in the paint shop could have been prepared and released for such a mission. The need to retrieve extra camera equipment from the Philippine Air Depot at Nichols Field, south of Manila, would have prevented any other photo flights from taking off before about 11:00 a.m.

If General MacArthur had given his permission to launch the Takao raid, one might also expect that General Brereton would have ordered Major Emmett O'Donnell—in charge of the other sixteen B-17s that had been deployed to Del Monte Field on Mindanao for dispersal—to bring those planes back to Clark Field immediately, for use in a second strike at Formosa that evening. On receipt of a message at 7:00 a.m., the two squadrons under O'Donnell's supervision could have been ready to fly from Mindanao to Luzon by about 8:30 a.m. Such a plan would have resulted in all sixteen 14th Squadron and 93rd Squadron Fortresses arriving at Clark before noon. If this had been the case, any change in events conditioned by early discussions between MacArthur and Brereton would not have altered history—at least in the greater scheme of things. It would have been very likely that the 14th and 93rd Squadrons would have arrived at Clark only to have been destroyed on the ground, while most of the 28th and 30th Squadron planes (on their mission to Formosa) survived! Perhaps ultimate outcomes are not so easy to reverse after all.

So, what may have been the fortune of the B-17s flying on the proposed Takao raid?

If the Fortresses had arrived over Takao at 10:00 a.m., there was a good chance that they would have found the target under partly cloudy, clearing skies. Distressingly, there would not have been many ships in port to bomb. The vast majority of Japanese invasion forces not already at sea were marshalling in the Pescadores Islands, nearly a hundred miles northwest of Takao. Whatever the Americans found as targets of opportunity around the harbor would have been subject to a crushing downpour of up to 120 600-pound bombs. It is true that the important Japanese dockyard facilities at Takao could have been heavily damaged, but their loss would not have had much impact on the upcoming campaign for the Philippines. If subsequent B-17 performance statistics held true on December 8, one of the Fortresses would have probably aborted the mission before reaching the target (due to engine or electrical problems), and two of them would have been plagued by bomb-rack release failures. Only a dozen would have been able to drop bombs. The mission would hardly have been the resounding success that the FEAF staff hoped to realize.

Some seventy-two Imperial Japanese Army Air Force (IJAAF) Nakajima Ki-27 fighters that could have risen to challenge the incoming B-17s—from airfields at Chaochou and Hengchun—would have had some difficulty climbing to the probable bombing altitude of 25,000 feet in time to catch the speeding Boeings. Even if these IJAAF fighters did attack, the damage their 7.7-mm machine guns would have done to a formation of well-armed B-17s at high altitude would have been minimal. The ninety Mitsubishi A6M Zeros that could have caused real trouble for the Flying Fortresses would have still been farther north, preparing to get into formation with the fog-bound IJN bomber strike force—which probably would have passed beneath the B-17 formation somewhere near Takao. In this scenario, a few American bombers could have been damaged, but most, if not all, would have made it back to Clark Field.

If the Americans had arrived over Takao a half-hour later (which is even more likely), they would have run directly into formations of Zeros heading south to escort the bombers. Undoubtedly, a large quantity of the A6Ms would have been quickly detached from their escort duties to effect an interception of the B-17s. The 19th Group formation might have even been broken up before it could do much damage to the port. In that scenario, one could expect a number of B-17s would have been shot down, or would have failed to return to Clark because of battle damage. Hypothetically, in combination with the losses that *could have* occurred among the B-17s that *might have* been moved to Clark from Del Monte, General Brereton's air force may have emerged from December 8 in even worse shape than it actually did! Might it have been a "blessing in disguise" that Lewis Brereton and Douglas MacArthur were unable to confer in a timely manner on that fateful day?

With reasonable assurance, one can expect that the returning B-17 formation would have found their all-important home base in a shambles. It is unlikely that the IJN raid on Clark Field would have been significantly deterred by the three P-40 squadrons, which may, or may not, have been properly vectored to intercept the bombers. Of course, the devastation of Clark Field itself, along with the destruction of the radar set at Iba Field in a concurrent IJN attack on that base, were the real tragedies of December 8. Any conceivable discussions between the generals would not have prevented those losses, and without those critical assets the air campaign for the Philippines was indeed "doomed at the start."

Options, Options

The Americans were not the only ones overwhelmed during the first few months of the war.

Three days after Pearl Harbor, Japanese planes sank the British battleship *Prince of Wales* and battle cruiser *Repulse*. Singapore surrendered on February 15. On that same day, the Japanese captured Palembang on Sumatra and with it a large portion of the Dutch East Indies oil reserves. On February 28, Japan crushed American, British, and Dutch warships in the Battle of the Java Sea, sinking two light cruisers and three destroyers.[9] Java, another major source of oil, fell on March 9.[10]

In April, a move by the Imperial Japanese Navy into the Indian Ocean led to carrier strikes against Colombo, the sinking of two British heavy cruisers, and a near-collision with the British Eastern Fleet, which could have been disastrous for Britain. Several days later, the IJN hit Ceylon (now Sri Lanka) again, raiding Trincomalee and sinking the British light carrier *Hermes*.[11]

From December 7, 1941, through May 6, 1942, Japan had run wild. At this stage, some Japanese might have favored switching to a defensive posture, but others, like Yamamoto, believed that the only way to avoid a long, drawn-out conflict with the United States was to remain aggressive.[12]

Where then should the next blow fall? The Japanese army wanted no part of an invasion of Australia, and another drive into the Indian Ocean to link up with Nazi Germany did not appear to be feasible, at least for the time being.[13] Should Japan advance on Fiji and Samoa or was it time to seek a showdown with America's aircraft carriers?

Q: *What was Japan's best move in the spring of 1942?*

John Lundstrom

Japan's most viable option was to go into the Indian Ocean and combine with the Germans to try to completely secure that flank, because the British were extremely vulnerable in the Indian Ocean. That would have required much closer cooperation with the Germans. If the Axis powers had been able to come up with the combined strategy that the Allies had, I believe the Japanese might have focused more on the Indian Ocean.

Going for Port Moresby was a good idea. The Japanese realized that they couldn't conquer Australia, but by taking Port Moresby, the Japanese would have further strengthened their defensive perimeter, just as they would have done by heading east and trying to grab Samoa and Fiji. All of this was predicated on defeating the Americans in battle, by knocking out the Pacific Fleet. Yamamoto always felt badly that he had not been able to take out the U.S. carriers at Pearl Harbor. The idea was to punish the Americans enough so that they would concede Japanese gains. Without a doubt, Japanese ambitions grew after their success at Pearl Harbor. They never dreamed that they would have the kind of superiority in numbers they had in the spring of 1942. The Japanese were extremely confident in the spring of 1942. I think that fact is largely unappreciated. Yamamoto really felt he could defeat the U.S. fleet if he had a chance to take on the American carriers.

John Burton

Leaders in Tokyo had no idea that March 1942 would represent the high point of their war effort in the Pacific. In all respects, Japan enjoyed quantitative, qualitative, and geographic superiority over the Allies at that time. Of course, none of the Allied nations even faintly suspected that the Japanese blitzkrieg was about to come to a crashing halt, largely through mistakes of Japan's own making.

The campaign for Java and the Dutch East Indies closed with the surrender of the Dutch on March 9, 1942. To that point, Japanese war planners had made few, if any, mistakes. Japan's general officers executed their campaigns without blunder. Objectives set by the high command in Tokyo were met, and met with much lower combat losses than anticipated. The empire's only notable setbacks were with Admiral Sadamichi Kajioka's premature and under-resourced attempt to take Wake Island, and General Masaharu Homma's protracted reduction of American and Filipino resistance on the Bataan Peninsula. Kajioka got his second chance—ultimately succeeding with some help from Admiral Nagumo's 2nd Carrier Division. Homma was removed from his command of Japan's 14th Army for missing the planned schedule of conquest in the Philippines.

After Java, the two clear campaigns remaining for Japan were: 1) to sever Allied supply lines into China by concluding the occupation of Burma, and 2) to cut off American reinforcement of Australia. Japan's overriding strategic objectives were to conclude the land battle for China that had dragged on since 1937, and to ensure that Australia could not be used as a springboard for the counterinvasion of the recently captured Indonesian territory and its oil fields.

Coincident with the conclusion of its Dutch East Indies campaign, Japan occupied Rangoon—the key Burmese port in the Allied supply route to China. All that remained along the critical Allied reinforcement path to Chiang Kai-shek's Chinese Army were the roadways in northern Burma, and the Imperial Japanese Army was moving steadily forward to close them. Because the ongoing campaign in Burma was basically one of land and air forces, the Imperial Japanese Navy was free to unleash five of its six fleet carriers—*Akagi*, *Hiryu*, *Soryu*, *Shokaku* and *Zuikaku*—and four smaller carriers: *Ryujo*, *Shoho*, *Zuiho*, and *Taiyo*. (*Kaga* was on its way to Japan for overdue repairs to its engines.) In addition, the IJN could utilize its four substantial land-based air flotillas to support new operations anywhere from the Bay of Bengal to the southwest Pacific.

In March, two thorns still poked at the victorious Japanese flanks: the American carrier task forces built around *Lexington*, *Yorktown*, *Enterprise*, and the newly commissioned *Hornet*, and the British Royal Navy assets that had been built up in the Indian Ocean around the carriers *Formidable*, *Indomitable*, and *Hermes*, all originally earmarked to support the defense of Singapore. Interestingly, when faced with a strategic decision to address either the British threat in the Indian Ocean or the American threat in the Pacific, the IJN chose the former. As a direct result, the Japanese navy engaged itself in an irrelevant and costly battle that accomplished very little for the Japanese war effort. While Japan had reason to fear both Allied naval forces, a little careful thought might have led to a different conclusion regarding priorities.

Arguably, the Japanese knew that several of the American carriers were operating in the South Pacific, defending the line of islands between Oahu and New Guinea. They did not know that *Enterprise* was about to take time off in preparation for the Doolittle Raid on the Japanese Home Islands. *Hornet* had only just arrived at the U.S. West Coast. *Saratoga* was still at Puget Sound Naval Shipyard undergoing repairs from the torpedo hit she had taken in January. During the first week of March, *Yorktown* and *Lexington* approached Rabaul, only to discover that Japanese forces were landing at Lae, New Guinea. After a quick strike against the Lae invasion force, the two American carriers were forced to retire to Pearl Harbor for replenishment; they had just completed the longest carrier cruise in U.S. Navy history. For nearly a month, the IJN could have run amok in the southwest Pacific with no interference from the USN.

The IJN knew a lot less about the status of the British fleet in the Indian Ocean. Had Japanese admirals known the true nature of that status, it is unlikely that so much of Admiral Chuichi Nagumo's fleet would have been assigned to the area. *Hermes*, with only one squadron of Swordfish biplanes, was almost inconsequential. *Formidable*, a brand-new ship, was still training with its aviators. *Indomitable* was just recovering its regularly assigned aircraft and crews from Aden, after serving as an aircraft ferry for Singapore-bound Hurricanes. In mid-March, none of these British ships was ready to take on Japan's Kido Butai.

From the British point of view, it is hard to imagine that Admiral James Somerville would have used his fleet—however strong it appeared on paper—to conduct offensive operations against Japan. Although he had one fast, modern battleship, *Warspite*, Somerville's other four battleships were relics of the previous war. In any event, he was too short of destroyers and cruisers to screen his slower battleships and *Hermes*, should he have decided to aggressively contest Admiral Jisaburo Ozawa's dominance of the Andaman Sea and the Malacca Strait. One must recall that the Royal Navy was still smarting from its December 10, 1941, loss of *Prince of Wales* and *Repulse* to the land-based bombers of the Imperial Japanese Naval Air Service. As daring a reputation as Somerville had developed in the Mediterranean, it is unlikely he would have placed his battlewagons under an umbrella of Japanese air power, nearly one thousand miles away from his base on Ceylon. Japan's holdings in Burma, and its supply line to Rangoon, were therefore quite safe from any immediate interdiction by the Royal Navy.

Nevertheless, the IJN threw its entire weight of fleet carriers—plus *Ryujo*, which had been on duty with Admiral Ozawa's fleet—into a preemptive attack against British bases at Colombo and Trincomalee on the island of Ceylon. Tactically, this assault was a victory: IJN carrier aircraft sank *Hermes*, along with the cruisers *Dorsetshire* and *Cornwall*, and ravaged the airfield and aircraft at China Bay. Psychologically, the loss of these ships was a devastating reinforcement of the stinging lesson that had been administered to the Royal Navy at the hands of the IJN in December 1941.

From a strategy standpoint, however, one can assert that Admiral Nagumo's foray into the Indian Ocean was a disaster for Japan. Regardless of whatever debate such a statement might stir, the Ceylon diversion cost the IJN more than a month of activity by its Kido Butai, and significant losses among its best-trained aviators on *Akagi*, *Hiryu*, and *Soryu*—enough to force all three of those carriers to return to Japan for replacement aircrews and aircraft. Few could argue that this impact on the IJN air service was not felt at Coral Sea and Midway. The absence of three veteran carriers and their planes at the Battle of the Coral Sea, less than a month later, directly facilitated an American strategic victory in thwarting Japan's invasion plans for Port Moresby, New Guinea. With a strange domino effect, the short-handed Japanese carrier air units at Coral Sea suffered heavy losses that kept *Shokaku* and *Zuikaku* out of the critical battle at Midway in June 1942. The resultant absence of those ships and planes was telling. The best intelligence the U.S. Navy could muster would not have thwarted overwhelming force—which Japan should have possessed if all six of its fleet carriers had been present at Midway. As it was, Midway was a considerably lucky victory for the United States.

By attacking Ceylon, the IJN missed its greatest opportunity of the war. The delay to Japan's Pacific War timetable conditioned by this unnecessary exercise cost Japan its best opportunity to decisively sever the most important Allied supply route to Australia. A solid Japanese grasp on Port Moresby, and perhaps the island of Efate, might have set the Allies back more than a year in their effort to strike back toward Japan's Home Islands. From an armchair admiral's retrospective view, that would have been the best strategic choice Japan could have made in the spring of 1942.

Frank Snyder

Although, in the spring of 1942, the Japanese were in a position to choose among three strategic moves—to the west (Ceylon), to the south (Australia), or to the east (Midway)—the Japanese army was generally opposed to any strategic move that might have lessened its ability to deal effectively with Japan's *real* enemy, Russia, with whom it had already been at war in the twentieth century. From the point of view of Axis strategy, a move toward Ceylon would probably have been their best choice, but from a strictly Japanese viewpoint, the best strategy would have been one that threatened Australia and tied down American naval forces (who were trying to maintain a line of communications from the U.S. West Coast to Australia), and would have effectively prevented U.S. forces from "rolling back" the extensive Japanese conquests of the first six months of the war in the Pacific.

In hindsight, it is easy to see that the Japanese should have avoided offensive operations to their east, particularly operations beyond the air-search radius of search planes based on the Pacific islands then under Japanese control. Anything beyond the search radius from Wake Island should have been avoided.

Douglas V. Smith

One needs to ask another fundamental question and consider the situation in late 1941 to answer this question. The question is, "Could you see any way for Japan to achieve her political objectives in World War II without Germany also being triumphant or successful in the war?" My answer to that question is, unequivocally, "No!" I can see no possible way for Japan to be successful in World War II without Germany also being successful. Then comes the situation: On December 6, 1941, the Germans were within twenty-three kilometers of the Kremlin and the spires of St. Basil's Cathedral. Small German tactical units were running through the underground subway tunnels of Moscow. The next morning, without having been previously informed by their Japanese ally, the Germans found out that the Japanese had attacked Pearl Harbor. The question then becomes, "What strategy should the Japanese have employed in 1942 that, having already brought the United States into the war, would have best served to ensure a German victory in Europe?" Also, the dominant continental focus of the Imperial Japanese Army would have to be taken into account. So also would the industrial capacity of the United States have to be considered. The answer then becomes simple. Almost nothing that the Japanese could have done would have achieved their strategic objectives. The Japanese navy wanted to precipitate an end to the war, by agreement, by the end of 1942. That was simply not going to happen after Pearl Harbor—nor was the Japanese army disposed to an early termination of the conflict. Thus—and this would have been very bad strategy—the only realistic option for Japan would have been to go on the defensive against America in the east and launch a major offensive in the Indian Ocean to link up with the Germans in the hope of taking India, the crown jewel of the British Empire. This would have provided, if the Germans were successful in the Caucasus, a bargaining chip with Britain, as well as denying oil to the British navy and making the Baku oil fields available to the German war cause. All of this would have had to be done while also securing the resources—particularly oil—of the former Dutch and British holdings in the Southern Resource Area for the Japanese war effort. Operation Orient between the Germans and the Japanese was intended to do just that. Unfortunately for the Japanese, the Germans were not inclined to give any substantial effort to Operation Orient. Thus, the prospects for such a sub-optimized strategy were grim.

A more realistic Japanese strategy would require going back in time. Japan—and particularly the Imperial Japanese Army with its continental focus—needed resources desperately from the Southern Resource Area. Japan had several realistic strategic options prior to its decision to attack Pearl Harbor. The Japanese should have been aware of the heightened isolationist spirit in America, both within the population and in Congress. Certainly, the American reaction to the earlier *Panay* incident should have made them aware of this. Thus, a Japanese initial move against the Dutch oil holdings in Southeast Asia would have been a realistic option that would have been unlikely to bring America into the war. Even extending such a move against the British holdings in this area would probably not have been sufficient to precipitate a declaration of war by the Americans. Moreover, notwithstanding the nonaggression treaty with the USSR precipitated by Adolf Hitler in April 1941, a move to sever the Trans-Siberian Railway would have greatly aided the German move against Moscow in December 1941. Sure, the Nomanhan defeat by General Georgy Zukov stilled any desire by the Japanese army to take on the Soviets again. Yet commando-type raids on the Trans-Siberian Railway could have prevented the Soviets from moving forty-odd cold-weather-capable divisions west for the defense of Moscow. If the Germans could have consequently taken Moscow, the Soviets would have lost their rail hub for their entire European theater. If—and this is a big "if"—the Germans could have held Moscow for a year or more, Stalin would in all likelihood have been forced into a negotiated peace on Germany's eastern front—especially if Hitler's late 1942 move toward the oil resources of the Caucasus had been successful, which it might well have been with a concurrent Japanese move against India. Thereafter, with Germany successful in Europe, something to trade for peace with Britain, and the American prospect of a Nazi and Fascist totalitarian Europe to contend with, Japan's freedom to its east would have been much less encumbered. Japan was overconfident and missed what little strategic maneuverability it had; yet it did have strategic options, which at least presented a hope of ultimate political success.

H. P. Willmott

With or without Hawaii, the Japanese, though they could not have guessed the reality in 1941, were saddled with an impossible problem in a war with the United States, and this reflected the wider problem of the war, not just the Hawaiian Islands and central Pacific. The simple fact was that by 1943 the United States was possessed of such massive material advantage that Japanese moral advantage—a willingness to die as the means of resistance—was increasingly at a discount. By 1943, the U.S. advantage was such that it was simply going to overwhelm Japan, and it is very difficult to see how Japan might have been able to shape things to a different end. In an obvious sense, Japan needed peace in May 1942 but, of course, there was no way in which the United States would have been prepared to accept a compromise peace. It may be that had Japan made her main effort in April 1942 in the Indian Ocean, and secured Ceylon and Aden, then perhaps there might have been a collapse of the British position in India and throughout the Middle East. London did believe that the Japanese raid on Ceylon in April 1942 might well herald the beginning of the end, and if this had happened then certainly very considerable changes would have been set in motion, but to what end is hard to foresee. The route to the Soviet Union via the Persian Gulf, which at this time was the most important sea route to the USSR, would have been closed, but to what end is difficult to determine:

Soviet survival in both 1941 and 1942 was achieved with very little assistance from Britain and the United States, and most certainly supply via this route was not of crucial importance at this stage of proceedings—and we tend to forget how important the North Pacific route, with Soviet shipping, was to be between 1943 and 1945.

But the point is that if the Japanese had undertaken a major offensive in the Indian Ocean and into the Middle East they would still have been forced to return to the central Pacific, and it is hard to escape the conclusion that at some stage or another they would have fought and lost a Midway somewhere. Thereafter the defeats would have followed in unbroken sequence in a war that really has three parts in terms of the Pacific—namely the naval, the maritime, and the political. The point, of course, is that Japan was beaten in all three—the war fought by fleets for bases and possession, the campaign fought by the United States against Japanese shipping, and the struggle for allegiance that Japan lost throughout eastern and southeast Asia.

Summary

In less than six months, Japan had seized more than a million square miles of Southeast Asia.[14] The Japanese had the resources they craved and the Imperial Japanese Navy had yet to taste defeat.

Now Yamamoto had to figure out how to solidify the Greater East Asia Co-Prosperity Sphere. Rather than wage a defensive war, he chose to move south against Australia, Fiji, and New Caledonia, and east for a showdown with the U.S. Navy.

Outnumbered in several categories, the U.S. Pacific Fleet seemed unlikely to stop the IJN.

Five

TURNING THE TIDE: SPRING 1942

BY THE SPRING OF 1942, Japan's options appeared limitless. A five-month rampage had vastly extended the Japanese empire, perhaps overextended it, and now the Americans fought back.

In February and March, U.S. carriers raided Japanese-held islands in the Marshalls and Gilberts. On March 4, the *Enterprise* ventured within 1,100 miles of Tokyo to strike Marcus Island. Then, in April, the Americans made their first appearance in the skies above Japan.

The Doolittle Raid

On April 18, a task force built around *Enterprise* and *Hornet* crept toward Japan. On board *Hornet* were sixteen B-25 Army bombers.

When enemy picket boats spotted the Americans about seven hundred miles from the Japanese coast, the bombers, led by Lieutenant Colonel James Doolittle, took to the air earlier than anticipated. The B-25s showered Tokyo and four other Japanese cities with bombs, and although damage was light, the attack showed the Japanese that their homeland was not invulnerable.[1] The Doolittle Raid gave Americans a morale boost, forced the Japanese to divert resources to strengthen their defenses, and made the elimination of America's aircraft carriers a top priority.[2]

Less than three weeks after the raid, two U.S. carriers, *Lexington* and *Yorktown*, battled the IJN to a standstill in the Coral Sea. Thanks to their participation in the Doolittle Raid, *Enterprise* and *Hornet* narrowly missed history's first battle between aircraft carriers, being about twenty-four hours away when the main action took place on May 8.[3]

Q: *What if the Doolittle Raid had not occurred and* Hornet *and* Enterprise *had been available for action at Coral Sea?*

John Lundstrom

If there had been no Doolittle Raid, then the *Enterprise* and the *Hornet* would have been sent south in late March or early April to the Coral Sea. The *Enterprise* would have relieved the *Yorktown* and there are indications that Admiral John S. McCain Sr. would have received the *Hornet* task force. That was Nimitz's preference. McCain would have joined Halsey in the Coral Sea. There could have been four American carriers in the Coral Sea at the beginning of the battle, after the *Lexington* finished a refit and the *Yorktown* spent a little time in port. They certainly would not have operated together had all four been available. I don't believe Halsey would have done that. He may have operated two separate task forces, each with two carriers, and that might have allowed the Americans to achieve a big victory at the Battle of the Coral Sea. Nimitz himself was very dubious about the Doolittle Raid, since it tied up half of the U.S. fleet's offensive resources for about six weeks. It left the Americans pretty weak trying to defend bases in the South Pacific. From the middle of March to the first of May, you essentially only had the *Yorktown* available in the South Pacific. That's a pretty weak defense and I know Admiral King was quite worried about it.

Frank Snyder

If there had been no Doolittle Raid, it seems likely that *Hornet* might not have been sent to the Pacific, since of the seven usable U.S. attack aircraft carriers, four were already in the Pacific, although *Saratoga* was off-line, being repaired at Bremerton. As it turned out, the Doolittle Raid effectively prevented *Enterprise* and *Hornet* from reaching the Coral Sea in time for the battle there, but it did ensure that *Hornet* would be available for the Battle of Midway.

If *Enterprise* had arrived in the Coral Sea in time to participate in the battle, this would have meant that Halsey would have been the officer in tactical command there, and he might have fought the battle differently. Furthermore, while *Enterprise* was sailing to join *Lexington* and *Yorktown*, Halsey might have discovered that the two large Japanese carriers that were in distant support of the Japanese operation to occupy Port Moresby were approaching via a route east of the Solomons—something that was unknown to Fletcher—and would enter the Coral Sea from the east, *behind* the two other U.S. carriers. Under those circumstances, Halsey might have declined battle until all three U.S. carriers were in company with each other, *or* he might have (with only one carrier) engaged the two Japanese carriers, and might well have lost *Enterprise*.

Peter C. Smith
The Doolittle Raid may have helped confirm Yamamoto's need for the final battle, but that had already been firmly decided upon anyway. The raid reinforced the acknowledged requirement, and had the bonus of bringing about the hitherto reluctant Japanese army on board. Unfortunately for the Japanese navy, the army's payoff was assigning *Shokaku* and *Zuikaku* to the Coral Sea operation, thereby depriving the Kido Butai of one-third of its strength, albeit by the two carriers considered the least efficient by the Japanese themselves. Had *Hornet* and *Enterprise* been able to reinforce *Yorktown* and *Lexington*, then a battle nowadays considered a "draw or tie" could possibly have turned into an American victory, always providing they were handled aggressively and not held back, of course. On the other hand, more U.S. carriers in the area might have led to more Japanese firm sightings of U.S. carriers and therefore more positive attacks and possible losses. So although *Shokaku* and *Zuikaku* were subsequently unavailable for Midway, consider the scenario had *Hornet* or *Enterprise* been heavily hit, or even sunk—what then for the Americans at Midway? Because Nagumo's four premier carriers were *always* going to be on the scene anyway!

Stephen D. Regan
One can argue that the Doolittle Raid served minimal purposes and actually cost far more than it was worth. Personally, I disagree with that assessment, but the availability of *Hornet* and *Enterprise* arouses some fun daydreams.

First, William Halsey would have been in command at Coral Sea instead of Frank Jack Fletcher. No one can doubt that Halsey would have lusted after Japanese carriers and would have attacked, attacked, attacked until either his four carriers were sunk or the Japanese fleet destroyed. I believe that he would have ordered his fliers to commence a night attack on the Japanese although our pilots were not trained for night fighting. With the additional two carriers in the Coral Sea, the various components of the Japanese fleet would have been discovered and a showdown would have been had.

The U.S. Navy would have spilled a ton of Japanese and American blood on the waters of the placid Coral Sea; however, the victor of that battle probably would have been Halsey. If the bloodbath that is envisioned had occurred, Japan would have had little with which to continue the naval war. Midway would never have occurred because I believe that Halsey would have sunk an additional two carriers at Coral Sea. Guadalcanal would have been the only other option for the Japanese, and we know the result of that battle.

It is quite possible that Japan's Asian policy would have turned toward Southeast Asia and the islands of the Pacific would have been ignored. A significant victory at Coral Sea would have changed the entire complexion of World War II.

Coral Sea

By early May, the Japanese were ready to extend their empire yet again, this time hoping to threaten Australia by taking Port Moresby, an Allied air base on the southeastern tip of New Guinea. If successful, Japan would menace New Caledonia, Fiji, Samoa, and the line of communications between Hawaii and Australia.[4] However, American code-breakers sniffed out the plans, giving the U.S. Navy a chance to strike a blow at the Battle of the Coral Sea.[5]

Nimitz sent two carriers, *Lexington* and *Yorktown*, to oppose the invasion, and on May 4, *Yorktown*'s planes announced their presence by peppering Japanese ships that were assisting with the invasion of Tulagi in the Solomons. Two days later, Rear Admiral Frank Jack Fletcher, in overall command, consolidated *Yorktown* and *Lexington* into a single unit, Task Force 17.

Now each side began to search for the other, both fully aware that whoever found their opponent first would have a big advantage in the upcoming battle, the first in which opposing fleets never saw or directly fired upon one another.

At first, Allied searches drew a blank. The Japanese carrier strike force, containing *Shokaku* and *Zuikaku*, with Vice Admiral Takeo Takagi in overall command, also scoured the sea to no avail. Faulty contact reports plagued both sides. In fact, on May 6, the Japanese unknowingly closed to within seventy miles of Task Force 17, but Rear Admiral Chuichi Hara, leading the IJN's 5th Carrier Division, opted not to launch an air search.[6]

Finally, on May 7, the Americans sank the light carrier *Shoho*, part of a separate covering force, the first Japanese carrier sunk by the United States, while the Japanese damaged the tanker *Neosho* and sank the destroyer *Sims*.

The battle climaxed on May 8, when American planes disabled *Shokaku* and ravaged *Zuikaku*'s air group, while the Japanese damaged *Yorktown* and sank the *Lexington*.[7]

Despite gaining a tactical victory, Takagi regrouped his battered forces to the north and Vice Admiral Shigeyoshi Inoue, in overall command of the campaign, postponed the Port Moresby invasion.[8]

Ordered by Yamamoto to resume the battle, Inoue instructed Takagi to pursue the enemy, but Takagi vacated the area for good after failing to make contact. Combined Fleet's eventual postponement of the Port Moresby operation handed the United States a strategic victory.[9]

Q: *How would Coral Sea have turned out if either side had discovered the other before May 8?*

Frank Snyder

Whenever one side would have learned of the position of the other, they would have launched a daylight strike, and the first effective strike delivered would probably have been decisive.

On May 7, the Japanese became aware that the Americans were west of their position, and, in late afternoon, they did launch a strike, but their aircraft failed to find the American carriers, so the Japanese planes returned toward their own carriers, and jettisoned their ordnance. Later, while still returning to their aircraft carriers, the Japanese planes sighted the American ships, but then had no bombs to drop. Although now alerted, the Americans did not launch a strike because there was too little daylight left (and in early 1942, aircraft did not normally land on aircraft carriers during darkness).

One clue about how much each side knew of the location of their enemy can be inferred from the coverage of the searches that each side launched at dawn the next day, May 8. The Japanese searched only the sector to their south, while the Americans searched in all directions. Each side's search planes located the other's ships at about the same time, and the attacks by each side against the other were made about the same time.

Peter C. Smith

A much fiercer and closely fought battle would have resulted, probably with heavier casualties, especially on the American side as they fought piecemeal as opposed to the concentrated force of the Japanese. *Shokaku* and *Zuikaku* might have been more heavily damaged, one even sunk, but American losses would also have been higher as they would have been more exposed to more experienced Japanese attacks and one damaged carrier, *Yorktown*, may well have been finished off.

Q: *Could* Lexington *have been saved?*

C. Peter Chen

After receiving torpedo and bomb hits during the battle, USS *Lexington* was rocked by an explosion that caused widespread fire. What took place, most likely, was that gasoline vapor had leaked into the generator room, and an electrical spark ignited the vapor. Flames were sent through vent systems to various parts of the ship. With no pressure in the forward water main and several telephone lines down, the damage control party was at a terrible disadvantage. About an hour and half later, the returning American pilots reported that one of the Japanese fleet carriers had been seriously damaged while the second was unharmed. *Lexington*'s damage control parties generally gave up by about 4:30 to 4:45 p.m., and the abandon-ship order was given shortly after 5:00 p.m.

Fast-forward three years: in March 1945, USS *Franklin* was challenged in a similar manner. Gasoline vapor explosion introduced fires, and the damage control efforts were hampered by damaged water mains and damaged communications systems. One major difference was that *Franklin* was saved. This presented the small chance that *Lexington* could have been saved as well.

In terms of the decision-making process, however, the environment at Coral Sea and that experienced by *Franklin* were vastly different. At Coral Sea, the Pacific War had just begun, and the Japanese navy was at the height of its power. Although the Americans knew that one of the opposing carriers was damaged and knew that the other must be busy recovering aircraft from her

USS Lexington *on fire and smoking at the Battle of the Coral Sea.*

damaged peer, the waters could be lurking with Japanese submarines, while it was unknown whether the Japanese could muster a task force to conduct a follow-up attack, both of which would expose those on *Lexington* to even further danger, especially as the very attempt to continue to save the carrier would be a gamble.

If she could have been saved, however, the damage would likely have been so extensive that it would have taken an extended amount of time to rebuild her. A typical carrier during this period took about a year to construct, plus time to put the re-launched ship through trials, equip her, and then to train the crew. In *Lexington*'s case, additional time would have been required to salvage through the mangled steel and charred equipment. Following this formula, the soonest she would have returned to the front lines would have been late 1943. While the Japanese navy was still a threat at this point, and the U.S. Navy certainly could have used her for the upcoming campaign in the Marshall Islands, the Americans had largely achieved dominance in the air by that time, and the addition of USS *Lexington* at that stage would somewhat have fallen victim to the rule of diminishing returns. On the other hand, given that the United States might have had the resources to devote additional manpower toward her repairs, perhaps *Lexington* could have been repaired at a quicker pace; in this case, the ongoing campaign in the Solomon Islands could have enjoyed an additional fleet carrier.

Q: *What if the Japanese had advanced against Port Moresby despite their losses on May 8?*

Lex McAulay

Takagi was in no condition to press on after May 8 because one carrier was sunk, one carrier was damaged and exiting the region, and airplane losses were severe, plus there was a fuel shortage (on both sides).

If the Japanese had pressed on with the invasion of Port Moresby after May 8, they would have done so against an Allied force that was alert and able to concentrate every element of offensive air power on a greatly reduced carrier force—*Zuikaku* only—in a relatively small area. The land-based bombers and fighters could have operated from a string of coastal airstrips from Townsville north, in the Torres Strait and at Moresby. One carrier could not have coped, and even supporting air strikes from Rabaul-Gasmata could not have overcome Allied air potential given the flexibility provided by the dispersed strips.

With everything from PBYs to B-17s, plus the U.S. Navy aircraft, *Zuikaku* would have been in mortal danger if it had been positioned to support the intended landing. Even the vulnerable Royal Australian Air Force (RAAF) Wirraways probably would have been sent against the transports and landing fleet, lugging 250-pound bombs. The Wirraway crews were warned to be prepared for such attacks, and the crews, believing they would be defending Australia, were aware that few would return if the bombs were to be delivered accurately.

Q: *If Japan had achieved a strategic victory in the Coral Sea and taken Port Moresby, what would have been the impact on Australia?*

Lex McAulay

Possession of the Port Moresby area would have allowed the Japanese to project naval and land-based air power along the northern and eastern coasts of Australia; probably as far south as Sydney. An Allied counterinvasion of New Guinea would have been extremely difficult to organize let alone execute, as there was only one railway along the east coast of Australia, north to south, easily cut at the many river-crossing bridges. The same applied to the one major north–south road, which also was very vulnerable to floods—the mountain ranges are close to the coast, with all that means in heavy rainfall quickly cutting the road.

To develop an invasion force using road and rail construction west of the coastal mountains would have meant a gigantic construction program, as nothing existed for north–south traffic inland. What there was existed east–west for export from coastal cities.

While the Great Barrier Reef would have been a rampart against IJN shore bombardment, given IJN superiority Australia would have needed a lot of coastal defense guns and air power covering channels through the reefs to make it an effective barrier.

Air strikes against Australia from Moresby would have resulted in a "Battle of Britain"–type campaign, which would have delayed further any Allied counterinvasion. If they thought it an advantage, the Japanese could have taken Horn Island, in Torres Strait off the northern tip of Cape York, where there was an air base, and used that in addition to Moresby bases.

An Allied counterinvasion of the Moresby area or elsewhere along the south coast of New Guinea would have been very expensive in casualties. Landing farther west would have been as bad as farther east, as at Milne Bay—swampy malarial terrain greater in extent than that on the north coast at Buna-Gona and extremely difficult to develop as a base for further operations to Moresby.

With Moresby and the south coast of New Guinea held, and Rabaul as the major base in the region untouched by air or naval efforts from Australia or U.S. Navy Pacific forces (apart from carrier strikes), the Japanese would have been in a very strong position in mid-1942 in the Southwest Pacific Area (SWPA).

At the Combined Chiefs of Staff level in Washington, D.C., I think the brutal reality would have been perceived: SWPA forces could defend Australia but would not be capable of a counterinvasion into southern New Guinea without enormous reinforcement in men, aircraft, naval forces, and every item of modern warfare for land, sea, and air operations. Therefore, the main counterattack against Japan would have been by the U.S. Navy forces. The Solomons campaign would have been fought as it was, probably with the same result, even if we accept carrier losses in the Coral Sea as they were—*Shoho* and *Lexington* lost, *Shokaku* and *Yorktown* damaged, and *Zuikaku* having lost her air component.

MacArthur would have been a defeated general in a defensive posture, though certainly agitating in Washington for a bigger role, and probably would have gone there to argue his case, maybe even for transfer to another theater—if FDR let him back into the United States.

The Philippines would have been bypassed, as the U.S. Navy intended, until MacArthur argued that it be liberated by U.S. forces to keep the promise made to the people there.

The Japanese had looked at the possibility of Allied attacks from Australia to the north and realized that they did not have to worry about movement north through central or western Australia, as it was mostly desert and undeveloped country with few useful resources and infrastructure and was incapable of supporting any major effort aimed at the Dutch East Indies. Any danger would come from an advance from eastern Australia to New Guinea and along the north coast of New Guinea through the East Indies to the Philippines or Malaya.

This did not eventuate because the Japanese failed to exploit the situation and turned away from success, as they did at Pearl Harbor, at Savo Island, with crucial convoys to New Guinea in 1942, and even as late as during the Battle of Leyte Gulf.

The Code-breakers

Despite the stalemate in the Coral Sea, Yamamoto forged ahead with plans to orchestrate a showdown with U.S. aircraft carriers by invading Midway, America's westernmost outpost in the Pacific, some 1,150 miles west of Honolulu. In Yamamoto's mind, Nimitz would almost certainly commit his carriers to save Midway.[10]

Blessed with superior numbers, Yamamoto could have kept everything simple. Instead, he frittered away his assets with a complex operation that involved over 150 vessels and an attack on the Aleutians to satisfy the Navy General Staff.[11]

The outnumbered Americans held the upper hand in one crucial area: intelligence. Cryptanalysts had broken the IJN's operational code, the JN 25 code.[12] While the Japanese eventually changed codebooks, Nimitz's intelligence experts were able to read IJN messages through May 28.[13]

By May 24, Nimitz had a good handle on where the Japanese were going, which ships were involved, and when they would arrive at their destination.[14] He decided to spring his own ambush by sending *Enterprise*, *Hornet*, and a hastily repaired *Yorktown* to Midway.[15]

Q: *If code-breakers had not uncovered Yamamoto's plans for Coral Sea and Midway, how would events have transpired in the spring of 1942?*

Jon Parshall

I personally don't view either Coral Sea or Midway as being "decisive" in the sense that losing them would have meant losing the war for the United States. The bottom line was that by mid-1943 the U.S. Navy was going to be larger than the Japanese navy. And by late 1944, it would be larger than all the other navies in the world *combined*. The Japanese could not affect that production equation one iota, as our shipyards might as well have been located on the moon as far as Japan's ability to attack them was concerned. No matter the outcome at Midway, even if it had been catastrophic for the United States, the Japanese could never match that productive output. In combination with the near impossibility of losing Hawaii, I can't conceive of us "losing" the war in the Pacific. Yes, losses at Midway and Coral Sea would have prolonged the war. But the Japanese, in the long term, could not cope with our productive dominance.

The other thing I'll say is that I find it hard to envision a set of circumstances wherein the United States would realistically have lost an unbroken string of naval battles to the Japanese. With all due respect to the tactical acumen and professionalism of the IJN, any serious study of the strategic and operational factors at work in the Pacific makes it abundantly clear that the United States was playing much better ball at the level of seaborne command. As the carrier operations prior to Midway as well as the battle planning for Midway itself make clear, the Japanese did not have a clear understanding of the strengths and weaknesses of the powerful naval aviation force they had created. Their willingness to divide their forces in the face of the enemy, their needless pursuit of secondary objectives (such as Operation Mo—Port Moresby) with those forces, their senseless preference for overly complex and distributed battle plans, their reflexive aggressiveness and willingness to expose their carriers to horrendous risks without the potential for commensurate payback, and their shocking callousness toward the expenditure of carriers, planes, and men all make it clear that this was not a navy that was well-positioned to prevail against the U.S. Navy. Simply put, the U.S. Navy was better led at the fleet level, was more adaptable, and was backed by an industrial and organizational powerhouse that the Japanese could never hope to match.

Peter C. Smith

Neither carrier battle would have taken place as they did, Port Moresby would have probably fallen, and an amphibious operation to re-take it would have had to be planned and carried out, with subsequent effects on Guadalcanal, because the U.S. Marines and the U.S. carriers could not be everywhere at the same time.

As for Midway, it would have probably have fallen, because Kido Butai would have been given ample time to erase Midway's air power at leisure and the Japanese heavy cruisers, reinforced by battleships if necessary, could have conducted unhindered bombardment of the islands' defenses at long range and reduced them to rubble and scrap-iron. The Japanese could then have occupied the atoll, landed aircraft, and got *all* units of Yamamoto's widespread force into ambush position, *before* the Americans committed forces to the area from Pearl. If, after the island had fallen, all available U.S. carriers had been rushed out (to atone for another Wake Island fiasco) they would have fallen exactly into the trap that Yamamoto envisaged, and would probably have suffered heavy losses.

Lew Hopkins

The plain fact of the matter is that if the American code-breakers had not broken the Japanese code, we wouldn't have known anything about Midway until it was over. The U.S. Navy and its carrier force were oriented toward the Southwest Pacific Area, so there's no question that the code was the tipping point for the defense of Midway.

Without the breaking of the code, I can envision the Japanese sailing against Midway and taking the island in a matter of a couple of days. I think at that point the U.S. Navy would have been forced to drop back and form a protective covering force for Hawaii.

If the Japanese had won the Battle of Midway, and the United States had lost all or nearly all of its carriers, we would have been forced to delay operations for a couple of years, trying to build up the necessary carrier strength to take on the Japanese. We would have had to wait for the carriers that were under construction. Without carriers to fight in the Pacific, you're stymied. I'm not sure the Japanese ever could have captured Hawaii, but the Americans certainly would have had to use all their resources to defend Hawaii, and there would have been no Guadalcanal in 1942.

Robert Swan

On Midway, we did not know exactly where the Japanese were going to attack. Thanks to the breaking of the Japanese codes, we were told the Japanese might attack several different areas, including Hawaii, Midway, the West Coast, and the Aleutians.

If Midway had fallen, it would have given the Japanese control of the Pacific and isolated Hawaii. It would have forced our industries to concentrate on providing much of the new war material—including new planes, good torpedoes, and bombs—to the U.S. West Coast and reduce the help we were giving to our Allies in Europe. I think we still would have beaten the Japanese but I'm not so sure we would have won the war in Europe.

Frank Snyder

If the Japanese plans to occupy Port Moresby had remained unknown to Admiral Nimitz, he probably would not have assembled in the Coral Sea the three cruisers, two aircraft carriers, and escorting destroyers to oppose the Japanese landing operation. Thus, it is likely that the Japanese would have been successful in taking Port Moresby.

If the code-breakers had not discovered the Japanese plans to mount a major effort to occupy Midway (*and* if the code-breakers had not previously been able to gain credibility by warning Nimitz of the Japanese plan to occupy Port Moresby), it is unlikely that Nimitz would have beefed up the defenses at Midway, nor would he have committed his three operational aircraft carriers that ambushed the carriers of the Japanese. It is likely that the Japanese would have been able to occupy Midway, *and*, if Nimitz had sailed his carriers to contest the Japanese occupation, the Japanese plan to lure the American aircraft carriers to their destruction might well have succeeded.

If the Japanese had won the carrier battle at Midway, the way would have been clear for them to execute plans to seize and occupy Midway; and if the Japanese had been successful in doing that, both of these outcomes would have drastically changed the course of the war in the Pacific.

The Americans would have had to reduce or abandon their efforts to keep open their line of communications to Australia, and to concentrate instead on dealing with the Japanese presence on Midway. The offensive operations in the Solomons in late 1942 might have had to take place without aircraft carriers, or to have been abandoned altogether.

If the Americans had been largely without fleet carriers in the second half of 1942, at least some of the U.S. submarine effort would have been diverted to interdict Japan's resupply and reinforcement of its forces on Midway. The U.S. strategists would have had to choose which one (or possibly two) of the following surface efforts would be undertaken, and which would have to be abandoned: defense of the U.S. West Coast, defense of the LOC to Alaska, defense of the LOC to Australia, and an offensive in the Solomons.

If the offensive operations in the Solomons had been undertaken, they would have needed to be supported by Army aircraft—in the absence of carrier-borne naval aircraft.

The launch of offensive operations across the central Pacific would have had to await the buildup of *Essex*-class aircraft carriers (as it did during the actual World War II), but with the infusion of some Army air, the offensive operations in the Solomons could probably have begun in late 1942 or early 1943—if such an operation had been chosen instead of, or in pursuit of, protection of the sea lines of communications to Australia or Alaska.

The victories over Germany and Japan both would have depended on the movement by sea of troops ever closer to those countries, and such movement would have depended on strategic decisions made in Washington about the allocation of landing craft. I think that the actual date of victory over Germany in May 1945 would probably not have been affected by the Japanese seizure of Midway, unless the reallocation of landing craft away from the European theater would have been made necessary by an invasion of our own at Midway.

Summary

By the spring of 1942, Yamamoto was in the driver's seat. However, what he needed most was a clear-cut objective that employed most, if not all, of his available resources. Whether it was Port Moresby, or Midway, or the elimination of the U.S. carriers, Yamamoto needed to direct all of his assets toward one task at a time. Instead, he attempted to accomplish all three goals in less than a month!

If not for American cryptanalysts, Yamamoto might have gotten away with it. The Imperial Japanese Navy may have gone unchallenged at Coral Sea and Midway. Instead, the code-breakers gave Nimitz the opportunity to send three of his carriers to Midway to surprise the unsuspecting IJN.

Six

DECISIVE BATTLE: MIDWAY I

YAMAMOTO HOPED TO MAKE Midway the penultimate victory, a smashing success that would force the Americans to sue for peace or at least give Japan enough time to consolidate her new empire.

Unfortunately for Japan, Yamamoto scattered his carrier strength. While two light carriers supported operations in the Aleutians some 1,800 miles north of Midway, the four fleet carriers of the 1st Carrier Striking Force, led again by Vice Admiral Chuichi Nagumo, was handed the job of softening up Midway for the landing of five thousand troops and support personnel. A line of submarines would patrol between Midway and Hawaii to alert Nagumo of the American response. Yamamoto's Main Force, with seven battleships, including Yamamoto's flagship *Yamato*, would sail three hundred miles behind the 1st Carrier Striking Force.[1]

American cryptanalysts uncovered just enough information to give Nimitz an opening. Task Force 16, built around *Enterprise* and *Hornet*, departed Pearl Harbor for Midway on May 28 with a new commander, Rear Admiral Raymond Spruance, a replacement for the ailing Vice Admiral William Halsey, who was laid up with a skin disease. Two days later, Task Force 17, with a patched-up *Yorktown*, left Pearl bound for Midway. Rear Admiral Frank Jack Fletcher, on board *Yorktown*, would be in overall tactical command of both task forces.[2]

Six Carriers

By the spring of 1942, Yamamoto's biggest advantage rested in his aircraft carriers. The same flight decks that wreaked havoc at Pearl Harbor—*Akagi*,

Kaga, *Hiryu*, *Soryu*, *Shokaku*, and *Zuikaku*—could potentially wipe out the remaining U.S. carriers if grouped together, as had been the case on December 7.

However, Coral Sea had cost Yamamoto the *Shokaku*, which underwent repairs for bomb hits, and the *Zuikaku*, which had to rebuild her air squadrons.[3] Two other carriers, *Junyo* and *Ryujo*, received tickets for the Aleutians, not Midway.

That left Nagumo with just four fleet carriers. Instead of delaying the Midway operation to allow *Zuikaku* to participate, as desired by the Navy General Staff, Yamamoto stuck with his original timetable, forcing Nagumo to make do with *Akagi*, *Kaga*, *Hiryu*, and *Soryu*.[4]

Q: *What if Yamamoto had delayed operations against Port Moresby and brought six fleet carriers to Midway, or had used all of his carriers against Port Moresby in May, shelving the Midway plan?*

Jon Parshall

If the Japanese had brought all six of their carriers to either Coral Sea or Midway, they probably would not have lost. The margin of victory at Midway, for instance, was slim indeed. With another two carriers in his fleet, Nagumo might have absorbed the losses of the morning of June 4 and still gone on to sink all his American counterparts. At Coral Sea, with six carriers to two American, the Japanese would almost certainly have walked away victorious and sunk both American carriers.

Anthony P. Tully

It is this option that provides the real key to a possible way out of the "Midway tunnel" that resulted from Yamamoto's fixation on the central Pacific thrust. In fact, in the writing of *Shattered Sword* an opinion already held was only further confirmed. The Japanese army had made a more correct assessment that the southwest Pacific offered the best hope of both forcing a decisive battle and perhaps some kind of political solution and compromise, gained by the leverage that a threat to Australia might bring.

Coral Sea, Operation Mo, partly misfired because it was treated as a "stepchild" operation by Yamamoto and the Combined Fleet. Even the fleet carriers assigned—5th Carrier Division—were the division judged the weakest link of the Kido Butai team that had rampaged across the seas from Pearl Harbor to the Indian Ocean. If Yamamoto had not insisted on Midway, or at least had allowed its postponement for several more weeks in favor of a maximum effort in the Coral Sea, the potential picture would have changed dramatically.

Homeward bound from the Indian Ocean operation, Nagumo's Kido Butai would have stopped over at Mako. From Mako, the *Zuikaku* and *Shokaku* would have detached and proceeded to the Battle of the Coral Sea, while Nagumo's other three carriers continued north for the homeland. In so doing, the latter would have arrived almost in time to intercept the U.S. carriers involved in the Doolittle operation, but for purposes of this speculation, this fact is important for another reason. Nagumo's itinerary shows that it was just as possible for *all* of Kido Butai to have proceeded to the Coral Sea, even as the Doolittle Raid attacked Japan. In other words, the full strength of Kido Butai would have arrived with killing force in the Coral Sea, with only *Yorktown* and *Lexington* able to oppose it at first, with *Enterprise* and *Hornet* to follow quickly.

Yamamoto and the Combined Fleet could have had their "decisive" carrier battle right there in the Coral Sea, in range of some Japanese air support and reconnaissance and with many of the advantages of surprise and concentration of force on their side. Furthermore, the Japanese army was almost certainly correct that the only remaining way after Pearl Harbor's miscarriage (politically) to obtain a political solution was some kind of threat to Australia as a bargaining chip. Failing that, strengthening the defense perimeter against the inevitable counteroffensive would be better served by an "expanded Coral Sea" plan than a Midway operation.

Peter C. Smith

Any delay of the Port Moresby attack would have given the Australians and Americans a much-needed breathing space to consolidate the New Guinea defenses and would have been counterproductive to the Japanese. There was no need to send six carriers to the Coral Sea and, with Midway already well along in planning, it would never have been contemplated. Don't forget: the whole idea of the Midway operation was to bring about the destruction of what was left of the U.S. Fleet, *not* the actual occupation of the island itself, something most postwar historians tend to overlook and which, indeed, seems to have been overlooked by many Japanese at the time. As in the Battle of Verdun in 1916, the *original* aim of the operation was lost once it was under way, and the obvious allure of acquiring real estate assumed more importance than the destruction of the enemy. In both battles, the secondary objective took over from the planned objective in the minds of the combatants once under way.

Nick Sarantakes
If the Japanese had brought six carriers to Midway, the battle would have had a significantly different ending. There was a good deal of happenstance that ended up shaping the course of this engagement. One of the most important was the arrival of U.S. dive-bombers over the Japanese carriers while their fighter coverage was out of position. Even then, the Americans needed even more luck—only ten of their bombs hit the carriers. The presence of another two carriers would have increased the number of planes available to the Japanese and made their defenses too strong for the Americans to penetrate. The most likely result of all from having these two ships would have been a modest Japanese victory at Midway. The Americans would still have lost the *Yorktown*—there is no reason to think otherwise—but these two additional carriers would have made no difference in determining the fate of the *Hornet* and *Enterprise*. Both would have survived, while all four—or rather six—Japanese carriers survived. The most important thing about this battle would not have been the loss of Midway Island—it would have been extremely difficult for the Japanese to hold it for long—but the fact that the Imperial Japanese Navy would not have lost four carriers' worth of mechanics and maintenance crews. The death of these highly trained and difficult to replace individuals contributed in significant, but indirect, ways to Japan's loss of air superiority as the war progressed. Less maintenance and upkeep reduced plane performance, which slowly but surely led to the loss of pilots. These individuals were even more difficult for the Japanese to replace given the rigorous nature of their flight schools.

Douglas V. Smith
If the Japanese had had the *Shokaku* and the *Zuikaku* available at Midway, they would have had an additional 144 carrier aircraft available. None of the hard choices that have been critiqued repeatedly in books would have confronted Vice Admiral Nagumo, as he would not have been placed in a position to weigh the trade-offs inherent in using his fighters to protect his strikes against

Midway, defend his carriers, and protect his torpedo planes and bombers en route to the U.S. carriers once they were located. Moreover, if the Japanese had concentrated on their Midway operation and had been successful in sinking all four U.S. carriers (because in this case *Lexington* would not have been sunk at Coral Sea and would have been available at Midway), they would have achieved their strategic objective de facto at Port Moresby. There would have been no U.S. carriers (assuming all were sunk at Midway) to contest a Japanese move against Port Moresby, and thus the Japanese objective there would have essentially been won at Midway. Ironically, by deciding to risk the *Shokaku* and the *Zuikaku* for a secondary strategic objective at Port Moresby the Japanese significantly raised the risk inherent in attempting to achieve their major strategic objective at Midway.

It is true that the ultimate outcome at Midway may well have been reversed if the Japanese had achieved their strategic objective in the Coral Sea battle. This would have put the United States on the strategic defensive until late 1943 when new-construction warships, including carriers, started entering fleet service, and with U.S. submarines' torpedoes failing to work properly until September 1943 there might well have been no choice on the American side but to come to some kind of negotiated peace while under attack throughout the Pacific. But one must consider that Admiral Nimitz probably would not have risked *Enterprise* and *Hornet* at Midway had he lost both *Yorktown* and *Lexington* at Coral Sea. This could have meant that the overall strategic situation—only two carriers in the Pacific until *Saratoga* returned to action in early June 1942—would have been exactly the same as after *Lexington* was lost at Coral Sea and *Yorktown* at Midway, except that the Japanese would have still had the four carriers of Kido Butai available.

Q: *Could the Japanese have scratched together an air group to allow* Zuikaku *to participate at Midway, or could the operation have been delayed to allow both* Zuikaku *and* Shokaku *to take part?*

Jon Parshall

As my research for *Shattered Sword* makes clear, yes, they could have. With a little scraping, *Zuikaku* probably could have put together an air wing of about sixty aircraft for the operation.

As for delaying the Midway operation: not unless the Japanese were willing to modify their overall operational schedule with regard to follow-on objectives in the Fijis and so on, which is just another way of saying, "Not unless the Japanese were willing to modify their entire strategic outlook on the war." Frankly, the reason they didn't have *Shokaku* and *Zuikaku*'s services at Midway was because their naval GHQ and Combined Fleet didn't have the mental discipline to assign rational levels of force to the target set in front of them. They were trying to gobble up too many objectives with too little force in too quick a time. Instead, they should have been doing fewer operations, but carrying a bigger stick to each of them.

Ronald W. Russell

Yes, but Parshall and Tully have shown that it would have required a fundamental change of doctrinal mind-set that the Japanese weren't going to undertake in early 1942. To the Japanese, the air group was an integral part of the ship. With *Zuikaku*'s air group decimated at Coral Sea, their only focus was to rebuild it. Taking squadrons from other ships would have been the same as trying to take the engines or guns from other ships. They didn't think that way, at least not then.

There were also some timing issues at Midway Island concerning the tides, the moon, and so forth. June 3–4 was optimum for an amphibious assault, or so the Japanese thought. As for *Zuikaku*, I don't know how much time was required to reconstitute its air group, and I think *Shokaku*'s serious damage at Coral Sea didn't get fixed until August.

Peter C. Smith

No, they had hoped to, but it was just not possible; there were insufficient trained aircrews on hand. According to one Japanese signal they even intended to add *Shokaku,* if she proved fit to fight on return to Japan, to the Northern Force, *not* to Kido Butai. In fact, her injuries proved to be much worse than they originally thought, and she almost capsized on the way home. By overestimating American losses, the Japanese naval high command considered Nagumo had sufficient strength to carry out *both* missions with just four carriers.

Due to conditions of moonlight for the planned bombardments, and tides for the invasion craft to get over the reefs, and other essential (but overlooked as not glamorous) considerations, the Japanese landing on Midway, if it was to take place, had to be done when it was done. Delay would have brought *Saratoga* into the scheme of things, and a long delay probably the *Wasp* as well. This would have negated the arrival of the patched-up *Shokaku* and the replenished *Zuikaku* with freshly trained air groups.

Surrendering Midway

In sending his carriers to Midway, Nimitz was risking precious assets. He later told Ray Spruance that while Midway was important, the U.S. carriers *were not* expendable.[5]

Q: *What if Nimitz had allowed Midway to fall?*

Jon Parshall

That would have been a perfectly reasonable decision on Nimitz's part. Midway was completely irrelevant to the defense of Hawaii, since it was outside the range of most land aircraft. Furthermore, it's my strong opinion (as stated previously) that the Japanese landing forces slated to attack the place didn't have nearly enough muscle to wrest the atoll from the Americans. (See *Shattered Sword* for a more detailed treatment of that subject.)

Even if the Japanese had managed to take the place, their ability to keep it in supply was marginal at best. It was exposed to submarine attacks against any ships bringing in supplies. Its own airfield was relatively small, and couldn't house a terribly large air group. Nor could those aircraft be adequately dispersed and sheltered from air attack. The Japanese themselves, as confirmed in a study done by their naval GHQ's Plans section in March 1942, didn't think they could hold the place. Nimitz, then, could have taken it back practically at will at any time the slender Japanese logistical thread looked like fraying.

Frank Snyder

Nimitz took two actions with respect to the defense of Midway: he beefed up the defenses at Midway, and he committed his three operational aircraft carriers, ostensibly to help defend Midway, but he actually positioned the carriers so that they would ambush the Japanese carrier force rather than interdict the initial Japanese air attack against Midway.

If he had taken neither action, Midway would have fallen to the Japanese. But if he had beefed up Midway itself, but had not committed his aircraft carriers, the outcome of the Japanese operation would have been in some doubt, and somewhat dependent on several other factors, including whether or not the fruits of the code-breaking were available, whether or not the U.S. submarines assembled off Midway were actively managed against the Japanese invasion fleets, whether or not the patrol torpedo (PT) boats added to the defenses of Midway were effective against an invasion force, and whether or not Japanese carrier aircraft could work effectively with Japanese ground troops.

The Japanese invasion of Midway might have succeeded, but it could have failed.

Repairing *Yorktown*

Damaged at Coral Sea, *Yorktown* limped home, arriving at Pearl on May 27. Normally, repairs might take months. However, in an amazing feat, some 1,400 workers made *Yorktown* battle-ready, and she sailed for Midway on May 30, significantly improving the Americans' odds of surprising Nagumo.[6]

Q: *What if* Yorktown *had not been repaired in time to fight at Midway?*

Jon Parshall

It's clear that Nimitz was willing to fight at Midway with only two carriers. Whether that would have been wise is another matter. However, if you look at it, *Hornet*'s air group contributed practically nothing to the actual combat operations on June 4. It's conceivable that a strike launched by *Hornet* and *Enterprise* alone could have achieved exactly what *Yorktown* and *Enterprise*'s aircraft in fact inflicted on the Japanese. However, it's also clear that the American margin for error would have been very, very slender indeed.

Barrett Tillman

The short version: we lose.

Yorktown, originally an Atlantic Fleet carrier, had rushed to the Pacific in December 1941 and proved invaluable over the next six months. She made a major contribution to blunting the Japanese drive toward Port Moresby, New Guinea, in the Battle of the Coral Sea in early May 1942. But in so doing she sustained bomb damage that threatened to sideline her for Midway. When she limped into Pearl on May 27, Navy yard workers estimated ninety days to repair her. Nimitz said, "I need this ship in three days"—and he got her. It was one of the minor miracles attending the battle.

With more combat experience than any other American carrier, *Yorktown* was essential to the defense of Midway. But her prewar air group was nearly exhausted, and had sustained losses at Coral Sea. Consequently, her Bombing 5 was kept on board as scouts, but *Saratoga*'s air group took up the rest of the slack. Though "*Sara*" had been out of action most of the year, her squadrons were fully capable of stepping into the breach, and they performed superbly in the battle. Bombing and Torpedo 3 arrived as intact units while Fighting 3 benefited from a wealth of experienced VF-42 pilots.

Beyond the squadrons, *Yorktown* herself and Task Force 17 were "all-up rounds." They had accumulated substantial experience under Rear Admiral Frank Jack Fletcher, and that corporate knowledge was evident off Midway.

"*Yorky*'s" younger sister *Enterprise* was the other old hand around the Pacific, though she had not fought enemy carriers. The "*Big E*'s" air group contained a wealth of talent, and performed extremely well at Midway.

Hornet did not.

Third and last ship of the *Yorktown* class, *Hornet* had been commissioned in October 1941. Her squadrons were "green as grass," having had no combat experience prior to Midway. She gained enormous fame for launching the Doolittle Raiders against Japan, but in doing so lost most chances of proper training for the air group. That was bad enough, but in addition, her captain, Marc Mitscher, selected vastly unqualified cronies to command the air group and fighter squadron, with results disastrous to the American cause. *Hornet* struck no blows against enemy carriers, lost her entire torpedo squadron and all the escorting fighters, and proved only marginally helpful otherwise. Her bombers and scouts were heard from later, in the Solomons.

Flying from *Yorktown*'s deck, *Saratoga*'s air group made a huge contribution to victory. Her squadrons conducted the only coordinated American attack of the battle, and her Dauntlesses sank *Soryu* unassisted and shared in the destruction of *Hiryu*. Her Wildcats fought the major fighter actions of June 4, downing more enemy planes than all other squadrons combined.

America probably would have won the battle without *Hornet* but would have lost decisively without *Yorktown*.

John Lundstrom

Without *Yorktown*, I believe Nimitz would have still sent Spruance to Midway with the *Hornet* and *Enterprise*. He was a very aggressive man. He was going to have Spruance sit on the Japanese flank and hit their carriers. Nimitz had confidence going into the battle. He had a lot of planes on Midway, in addition to his carrier planes, and he thought Midway's planes would do better than they did. He also had submarines on hand and thought they could make a contribution. Nimitz was very disappointed with the performance of the American submarines.

Without *Yorktown*, I believe Nimitz would have sent *Enterprise* and *Hornet* to raid the Japanese carriers and then pull back. He might have altered his tactics but he still would have contested the Japanese advance on Midway.

Frank Snyder

If *Yorktown* had not been present at Midway, two Japanese aircraft carriers would have survived the air strike by planes from *Enterprise* at 10:25 a.m. The absence of *Yorktown* might not have mattered if the planes from *Hornet* had participated in the attack on the Japanese carriers (instead of having been sent off in the direction of their position as originally reported). But even if the two squadrons of scouts/bombers from *Hornet* had participated in the air strike, the absence of *Yorktown* would have meant that the Americans would not have had any scout planes to search for the Japanese ships that had *not* been destroyed in the 10:25 a.m. strike. And if *Yorktown* had not been present, it is unclear whether Nimitz would have ordered the earlier air search to the northeast, or whether Spruance would have carried out such an order; and if he *had*, whether he would have waited, as Fletcher did, for the search planes to return before heading to his launch point.

Scouting the Enemy

Action off Midway began on June 3, the same day that the Japanese bombed Dutch Harbor in the Aleutians, as one of Midway's Catalina patrol planes spotted the IJN's Midway Invasion Force. The following day, at 4:30 in the morning, while the Catalinas searched for Japanese carriers, Nagumo launched his Midway strike force and scout planes.

As with any carrier clash, early detection was the key. At 5:34 a.m., a PBY (patrol bomber) struck gold, uncovering enemy carriers northwest of Midway.

At 6:30 a.m., Japanese bombers began to plaster Midway. The atoll's ragtag assortment of planes, including Avenger torpedo planes, B-26 bombers armed with torpedoes, Dauntless and Vindicator dive-bombers, and B-17s had already escaped to deliver multiple attacks on Nagumo's carriers.

Initially, Nagumo launched just seven scouts. *Kaga*, *Akagi*, and the battleship *Haruna* sent up a single scout plane each while four others catapulted into the air from cruisers *Chikuma* and *Tone*, a questionable number considering the vast amount of ocean to search.[7] Due to catapult problems, one of the cruisers, *Tone*, also ran into delays in launching a scout.

At 7:15 a.m., with no enemy ships in sight and with a call from his Midway strike leader for another attack on the atoll, Nagumo ordered his torpedo planes rearmed with contact bombs.[8] At 7:28 a.m., one of *Tone*'s search planes found the one thing that Nagumo had not expected: American ships two hundred miles away. By 8:30 a.m., *Tone*'s scout plane had also confirmed the presence of an enemy carrier.[9]

With a two-hour head start in sighting their opponent, Fletcher and Spruance had the drop on Nagumo. The Americans would also be hitting four Japanese carriers operating together.

Q: *What if the Japanese had done a better job of scouting at Midway, or had divided their carriers into separately operating units?*

Jon Parshall

Better scouting clearly would have aided the Japanese. The crucial time period for detection of the Americans was between 6:00 and 6:30 a.m., and had there been more search aircraft on the northeastern flank of Kido Butai, one or both of the American task forces might have been detected in time to launch strikes against them before their own aircraft were launched. If Nagumo could have actually launched a coordinated strike against the Americans, with a fully constituted combined arms air group, the Americans would potentially have been in very, very big trouble.

I personally don't know that operating the Japanese carriers in two separate task forces was the right answer for the Japanese. The problem with such an arrangement, of course, was coordinating their offensive activities (as the air operations of the American carriers makes all too clear). Concentration of the carriers also made it possible to cover the entire force with a smaller total combat air patrol (CAP). The problem was less one of concentration than of poor scouting and (frankly) luck on the part of the Americans. Japanese fleet air defense worked fairly well throughout the morning, when it was presented with a series of single-vector, single-altitude threats. Things fell apart when the Americans, through dint of sheer luck, presented the Japanese with a large-scale, multi-vector, multi-altitude threat.

Peter C. Smith

The Japanese flew all available scouting missions, given the number of cruiser-mounted aircraft they had, and could therefore not have improved on that. Faults in launching procedures could not have been foreseen. Those that went looked roughly in the right segments but did not report very well even when they did make contact. The American Midway-based aircraft made similar location errors of course.

Splitting the Japanese carriers might have saved them having all their eggs in one basket when the U.S. dive-bombers arrived; on the other hand, the combined CAP scored victory after victory against all the many U.S. waves of attackers *prior* to the arrival of Lieutenant Commander C. Wade McClusky and Lieutenant Commander Richard H. "Dick" Best et al. Radar would have made all the difference here.

Sumner Whitten

Japanese scouting during the battle was not that bad. The Japanese bosses, the battleship admirals, just didn't believe what their own two scout planes reported. Had the Japanese postponed their dumb Alaskan adventure and concentrated all of their carriers at Midway, it could have been disastrous for us. But separating their carriers into four different segments would have required command and control experience and communication equipment, which the Japanese did not have. It would have dispersed the Japanese carriers over too large of an area. However, the entire Japanese armada was commanded by a battleship admiral who did not understand diverse command. Besides, the whole deal was based on the December 7 Pearl Harbor attack philosophy. The

Japanese were not trained in dispersed carrier operations, did not have suitable command and control, and did not have the refueling capabilities for dispersed operations.

Frank Snyder

The two navies conducted scouting differently. The Americans relied on carrier planes to do their scouting while—at the time of the battle—the Japanese used "float" planes (seaplanes) from their cruisers and battleships. The advantage of using carrier planes for scouting was that they were aircraft of higher performance, and their radio communications would have been directly with an aircraft carrier, and would not have had to be relayed through their "home" cruiser or battleship. The disadvantage was that using carrier planes for scouting might have reduced the number of carrier planes for bombing. The Americans had solved this problem by having two squadrons whose pilots were trained for both scouting and bombing. Thus, American carriers had four squadrons each—two for bombing/scouting, one for fighter planes, and one for torpedo planes. The Japanese carriers had three squadrons each—one each for fighters, bombers, and torpedo planes. At the Battle of Midway, the Japanese were beginning to experiment with carrier planes for scouting, but during the battle, these planes suffered radio problems. So the Japanese could have "done a better job of scouting" by using carrier planes, but that was not then the Japanese navy's practice.

Some writers have suggested that the Japanese should have launched search planes in two "cycles," the first cycle taking off several hours before the launch of the attack planes, the second cycle, at the time that the attack planes launched. Two cycles might have been better than one, but one of the Japanese scouts must have flown over two American aircraft carriers without seeing them, and the scout that did report seeing them had been catapulted a half hour late, and had turned a half hour early to his cross-leg, and thus had entered the adjacent search area, previously covered by the search plane that had failed to sight the American ships.

The idea of dividing the four Japanese carriers into two separately operating segments would never have occurred to a navy that enjoyed great success with massive strikes that integrated planes from all carriers present, a navy that—like Mahan—believed that one should "never split a force."

Hal Friedman

This question relates to something that I have been speculating with for a bit, that is that the Japanese mistake was not that they did not divide their carriers but that they too greatly separated all of their forces. This was a mistake the Japanese made a number of times, especially at the Battle of Leyte Gulf. At Midway, I think the Japanese should have operated a more integrated and larger force, in effect, a combined arms naval task force. This would have entailed Admiral Yamamoto staying in Japan so as to be a focal point for command, control, and communications, much as Admiral Nimitz was in Hawaii. Next, the Aleutians operation should have been abandoned. As even Commander Fuchida remarked later, as did many historians, the operation made no sense since the objective of the Midway operation, in addition to capturing the island, was to lure the American fleet into a decisive battle, not lure it away to the north.

If Admiral Boshiro Hosagaya's forces, in the Aleutian Islands, had been added to Admiral Nagumo's—with Nagumo remaining as OTC (officer in tactical command)—Nagumo would have had that many more planes available to him, not to mention additional surface escorts. In addition, if Yamamoto had remained in Japan, the main body of the Combined Fleet could have been assigned to another officer, such as Admiral Matome Ugaki, and also placed under Nagumo's command. It might have even been possible to assign some of Admiral Nobutake Kondo's surface forces and aircraft carriers to Nagumo, further reinforcing the tip of the sword. The addition of all of those surface ships would have added to the antiaircraft strength of the Nagumo force. While Japanese antiaircraft left a great deal to be desired at Midway, more would have been better than less.

In addition, while the doctrine of using battleship and cruiser scout planes was not the best, if Nagumo had had more surface-ship scout planes available to him, he could have mounted a more effective search for the American carriers. If the main body's carriers had also been added to Nagumo, then those light carriers' air groups could have been employed as antisubmarine and combat air patrol forces, thus freeing Nagumo's carriers to focus entirely on strike warfare. Even if Nagumo had then lost his carriers to American strikes, he would have had a huge surface force that might have been able to more aggressively pursue and engage Spruance at night. The last point would have entailed Nagumo actually being more aggressive after the loss of his carriers, but it did the Japanese no good to have the main body three hundred miles behind the Nagumo force. The larger force would have presented an even more inviting target to Spruance the next day, but both of his remaining air groups were quite whittled down, as was Midway's air power. I think if the Japanese had had any hope of salvaging the operation, this was it. Here, I am envisioning what the Americans did later by integrating surface ships into the Fast Carrier Task Force.

Silence from Yamamoto

In the days leading up to Midway, the Japanese tried to pinpoint the whereabouts of the U.S. carriers. A special reconnaissance of Pearl Harbor, Operation K, was to have involved flying boats operating out of the Marshalls, but the presence of American ships at the refueling point, French Frigate Shoals, scrubbed the operation.[10]

By June 2, Yamamoto also had hoped to have a line of submarines in place between Hawaii and Midway to spot enemy warships headed toward Midway, but the subs failed to arrive on station until June 4. The delay turned out to be inconsequential, since at that point *Enterprise*, *Hornet*, and *Yorktown* were already safely located some 325 miles northeast of Midway.[11]

Beyond the cancellation of Operation K and the failure of his subs to spot opposing ships, a few other developments might have concerned Yamamoto. There was the recent increase in radio traffic that seemed to indicate that an American task force might be at sea. There were also reports that Midway's search planes were flying long-range patrols.[12]

However, to maintain radio silence, Yamamoto never informed Nagumo about Operation K and never sent any other information that he might have possessed that might have increased Nagumo's vigilance.

Q: *If Nagumo had received news of any warning signs that U.S. ships might be near Midway, would he have acted any differently? What if Operation K had been completed successfully?*

Jon Parshall

From a very strict interpretation, Nagumo actually had precisely that sort of information in his hands. It is clear from *Akagi*'s air group report that they were

Vice Admiral Chuichi Nagumo, commander of the Mobile Force at Midway.

aware *before the battle* that the air station on Midway was *not* operating like a sleepy backwater that had no idea as to its impending fate. Quite the contrary; it was a beehive of activity. If the enemy base appears to be on edge, is it not safe to at least plan for the contingency that the Americans have at least some inkling that something's up? And if one makes *that* assumption, it doesn't take much to make the follow-on assumption that *if* the Americans are edgy about Midway, that they might also have some carriers in the area? The question then becomes: what to do about it? At the very least, one can assume that Minoru Genda, who directed air operations for Nagumo, would have beefed up his reconnaissance assets. The Japanese might have altered their course, so as to approach the islands more carefully, after having done a more thorough recon of the northern approaches to the island. Bear in mind, though, that any such modifications would have to have taken into account the likely impact on the Midway landing schedule, which was very rigidly set.

Peter C. Smith

Various reports were available to Nagumo, but he ultimately chose to ignore his basic instruction and keep half his aircraft ready in case U.S. carriers turned up. The second-wave attack against Midway was unnecessary; the island's air forces hardly existed after the first attacks and could have been mopped up at any time, once the search planes had fully searched and reported. The risk of re-arming did not have to have been taken as early as it was, Midway was going nowhere and its sting had been drawn; it could have been polished off at leisure later in the day.

If Operation K had gone forward, that might have given the Japanese more pause for thought, but don't forget they believed *Yorktown* had been sunk, or was so badly damaged that, like *Shokaku*, would be out of combat for a long time. And an American signals ploy down in the Coral Sea made them believe that at least one, or maybe two, of any U.S. carriers still afloat was still in the southwestern area.

Stephen D. Regan

Nagumo was plagued with dual orders: attack Midway for occupation and defeat the U.S. carriers in an all-out battle. The orders came into diametrical opposition when the Japanese scout planes, which were to refuel on French Frigate Shoals, were unable to check on the carriers supposedly in Pearl Harbor. Assuming that the carriers were in Hawaii, Nagumo acted as he did. Had he known that the carriers were not in Hawaii, he certainly would have acted otherwise.

The entire story of the attack on Midway was about as expected. Nagumo did not anticipate Midway requiring a second wave, nor did he figure that his returning planes would arrive about the same time as his intelligence discovered the U.S. carriers. He was caught with his pants down.

If Nagumo had known about the carriers, he would probably have offered a more conservative attack on Midway while fully preparing to launch a torpedo attack on the American carriers. With a modest knowledge of the whereabouts of the American carriers, he would have clearly hammered them as they were hammering him. Probably *Soryu*, *Akagi*, and *Kaga* would have suffered significant damage, if not sunk. *Hiryu* would have escaped.

On the other side of the coin, *Yorktown* was doomed. She was barely seaworthy as it was, and she could handle little in additional injury. *Hornet* and *Enterprise* were in the same vicinity and an increased attack may very well have seriously damaged at least one of these priceless ships. (Remember that *Saratoga* was being repaired.)

In the end, had Nagumo's belatedly launched scout plane spotted the Americans earlier, or if the initial check on the carriers at Pearl Harbor had been successful, then Midway would have been something of a draw. Midway would not have been invaded because of the Japanese carrier damage. The Japanese would have continued as they ultimately did in the Solomons. America would have had second thoughts about the invasion of Guadalcanal. That is the biggest question!

If the Japanese had successfully controlled Guadalcanal, with a full-blown occupation complete with airstrip and appropriate land-based planes, King and Nimitz may have had some serious fights with General MacArthur. The legendary MacArthur may well have had the upper hand and demanded that the Navy protect Australia and New Zealand while using his Army Air Force to bomb away on Guadalcanal or to island-hop to Bougainville, or more likely take New Guinea.

I think that the taking of Guadalcanal, Tarawa, and even Bougainville would have taken place much later in the war. More interestingly, Marshall and MacArthur would have been the men in charge instead of Nimitz and King. The entire scope of the Pacific war would have changed. It is hard to imagine history without the Solomons campaign: possibly no Iwo Jima or Saipan, and a lengthier, Army-oriented war. In this scenario, Peleliu would have become significant instead of a stupid slaughter of men. Saipan and Guam may have been overlooked for a Formosa invasion.

C. Peter Chen

Usually, historians try to keep to the principle of ceteris paribus with "what if" scenarios. With the Battle of Midway, there were a lot of variables that, even if we only change one, could have changed everything. Operation K, the plan to send seaplanes to see whether the American carriers were in port, was aborted due to an unexpected American presence at the planned forward base. The submarine net that was supposed to catch the American carriers sailing north simply got in position far too late. The scout planes from carriers turned home too soon, launched too late, or perhaps simply did not look hard enough. The one element of the scouting plan that succeeded, submarine I-168 that reported Midway bustling with activity, for some reason caused so little alarm that no one thought it should be brought to Nagumo's attention.

What if Japanese submarines had been given enough priority and resources so that they had a chance to catch the American carriers? What if the scout planes had been more successful? Or, the "what if" that almost happened, what if someone had actually realized that Midway did not look like a base about to be caught by a surprise attack?

Knowing that the Americans were expecting an attack, Nagumo's search planes would have been on the lookout for the American carriers, rather than merely conducting routine patrols searching for ships they did not expect to be there. With Midway itself being downgraded to be a secondary target tactically, dive-bombers and torpedo bombers would have been armed and ready to launch from at least two of the four carriers, while the remaining one or two carriers would have maintained a tight combat air patrol above the fleet, fighting off the ineffective and uncoordinated earlier attacks on the task force. The Japanese would have found the American carriers shortly and dispatched them to the bottom of the sea. The American attack aircraft that had been dispatched prior to the Japanese finding their mother ships' locations might have found, damaged, and perhaps even sunk one of the Japanese aircraft carriers, but they would have found that their home flight decks were below the waves, and not all of them would have had the fuel needed to make it to Midway. Midway itself would have been taken after an intense bombardment and the subsequent landing, but long-term consequences would have been far more damaging for the United States. As the Japanese army was already complaining of overextension, a follow-up invasion of Pearl Harbor would have been rather unlikely, but the Americans would still have had plenty to worry about. With the fleet carriers lost, the Americans essentially would have fielded no significant carrier force in the Pacific until later in 1943. This continuation of Japanese success, rather than the change in momentum that real history offered, would have guaranteed a successful Japanese conquest of all of the Solomon Islands, and would have subjected Australia to regular air raids. The Pacific War, ceteris paribus, likely would have been won by the Americans ultimately, but the eventual victory probably would have been dragged on into at least 1946, and the events that might have taken place beyond that point would be impossible to imagine.

Balancing Resources

On the morning of June 4, Nagumo hit Midway with thirty-six Kates from *Hiryu* and *Soryu*, thirty-six Vals from *Akagi* and *Kaga*, and nine Zero fighters from each of his carriers.[13] As directed by Yamamoto, a 108-plane strike group was placed in reserve, including Kates armed with torpedoes to deal with any enemy ships that might be encountered.

However, the battle turned when Nagumo rearmed his second wave, switching from torpedoes to bombs on the Kates and replacing armor-piercing bombs on the Vals with fragmentation bombs, all for a second strike on Midway.[14] Mitsuo Fuchida would later claim that he had supported attacking Midway with planes from just two of the four carriers, holding the rest in reserve.[15]

Q: *What if Nagumo had hit Midway with squadrons from two carriers and kept the rest of his planes in reserve to deal with enemy ships?*

Jon Parshall

I personally think that this approach would have yielded even poorer results for Nagumo. The reason, historically, that each Japanese carrier launched a deck-load strike of half of their aircraft (typically between twenty-four and twenty-seven at a time) was that this number was the largest group of aircraft that could be conveniently spotted and launched on the two smaller flight decks (*Hiryu* and *Soryu*). If they used their entire air groups, they had to have multiple spotting cycles for a single launch. This deck-load approach paid admirable results, in that the Japanese could launch a coordinated strike from all four of their carriers, and get the aircraft formed up and on their way to a target within less than fifteen minutes. By contrast, the American carriers, which insisted on using larger launches with multiple spots, often took upwards of an hour to get their similarly sized strikes into the air, leaving the first aircraft to take off orbiting overhead and burning precious fuel (and hence strike range). This is precisely why some of the American launches during Midway were uncoordinated and scattershot.

Having the Japanese resort to using just two carriers against Midway, and holding two in reserve, would have exacted undue penalties of either force or time. If the Japanese had used only a deck-load strike from each of the two "Midway attack carriers," then they would have ended up hitting the island with only half the force that the actual morning strike of 108 aircraft used. Clearly, that was not a good idea. As it was, the 108-aircraft strike achieved only moderate results against the American bastion, and it was immediately clear that a second strike was needed. A force only half as large would have achieved far less, of course, and might well have taken proportionally even heavier casualties than the already very heavy casualties the 108-plane force was subjected to. It is a military truism that marginally sized forces often take outsized casualties in comparison to a larger unit. So, with a strike of, say, fifty-four aircraft, the odds were that the American defenses (particularly their fighters) would have faced a much smaller Japanese fighter escort, and thereby exacted even heavier casualties from the Japanese bombers than they historically did. If that is so, those same bombers would likely have done proportionately even less damage (having had more aircraft destroyed by the American CAP before they could deliver their payload against Eastern or Sand island). This doesn't seem like a good way to go.

The other option, using the full air groups of the two "Midway attack carriers," isn't good either. To do so, the Japanese would have been going against their normal deck-load spot doctrine, and would therefore have required a much longer spot/launch/spot/launch cycle in order to get the full strike up in the air. That means that they would have had to potentially launch from closer in to Midway (which would have placed them closer than they needed to be, and reduced their freedom of maneuver), and it would have tied up the flight decks for longer than was necessary, thereby decreasing the ability to launch CAPs from each of the carriers should the need arise. Either approach, to me, seems less efficient than using the normal deck-load strike routine with all four carriers that Nagumo historically employed.

Frank Snyder

If Nagumo *had* struck Midway with squadrons from two carriers, and had held planes from his other two carriers in reserve, the outcome of the battle would have been unchanged. Nagumo's plan—to strike Midway with bombing squadrons from two carriers, and torpedo squadrons from the other two carriers—had the advantage of using all four decks for launch and recovery of each strike, shortening the time required, and simplifying the process of spotting planes on deck in each carrier.

The shape of the battle resulted from Nagumo's failure to allocate any of his fighters for *defense* against attack. He sent half of his fighters with the Midway strike, and retained half of them in reserve to accompany a strike against American ships, if they were to show up. This Japanese tendency to maximize their offensive strikes came naturally from their experience in the early months of the Pacific war, and in their previous experience in using carrier planes to strike the Chinese. So, when American planes arrived (unexpectedly) from Midway to attack the Japanese carriers, the Japanese carriers launched the "reserved" fighters to defend (quite successfully) the Japanese force. As a result, when Nagumo received the report about the presence of American ships, there were (unfortunately for the Japanese) no fresh fighters to accompany a strike by the bombers and torpedo planes held in reserve. Nagumo had witnessed how easily his own defending fighters were able to shoot down American torpedo planes and bombers that arrived unescorted. He decided to delay his own strike on the Americans until fighters to escort such a strike could be made ready.

Alvin Kernan

Carrier-to-carrier battles were new to warfare in the Pacific War, the first having taken place in May at the Battle of the Coral Sea. But it did not take long to establish that the key to this kind of battle was to concentrate on the enemy carriers, ignoring everything else, and that the winner would always be the side that got to its enemy "fastest with the mostest." Scouting, response time, and the management of the flight deck became critical time factors, for whoever was in the air first had a tremendous advantage over their enemy.

It was at all these points that the Japanese fumbled at Midway. Nagumo's scouting was delayed, and when he heard that there were enemy ships, of whatever kind, within striking range on his eastern flank, he should at once have launched his available planes, already loaded for a naval battle, against them. He could have been in the air close to 8:00 a.m., the same time the Americans were departing to attack him. He would have discovered how powerful an enemy he was facing when his planes got to the target. It would have been easier for him to launch immediately had he kept his reserve attack force on two carriers, instead of distributing them among all four, for then he could have landed his returning Midway attack group while he was sending the new group off. Even with his reserve on all four carriers, he could have delayed the landing of his Midway strike until his naval attack group was in the air and on the way. Some of the Midway planes would have crashed from running out of gas or damage, but the loss would have been well worth it if he had obeyed the first cardinal rule of carrier warfare: get the biggest attack you can mount in the air as soon as possible once an enemy has been located.

Since the Americans would have been in the air at the same time the Japanese were, or even slightly earlier, the American attack could not have been prevented, but the Japanese strike could have arrived at the American fleet with more planes than it eventually did. The skill of the Japanese at this kind of carrier attack had already been tested and proved, and later in the same day small strikes against the *Yorktown* disabled that ship with bombs and two torpedoes. The Japanese first all-out strike would have sunk several if not all of the three American carriers and given the Japanese at least a draw in the battle. Furthermore, the damage they would have inflicted on the American carriers would have probably saved the fourth of their carriers, the *Hiryu*.

For all the strength and skill of the combatants, Midway involved, for both sides, a great deal of chance. But the deciding factor was the Japanese decision at 7:30 a.m. to delay their strike against the American fleet. Had they launched at once with all the power available they probably would have sunk two of the American carriers, possibly all three. In doing so they would have saved their fourth carrier, the *Hiryu*, and controlled the approaches to Midway, making it possible for their invasion fleet to land and their plans to be fulfilled. They might well have, as they planned, forced the American fleet to abandon Pearl Harbor and retreat to the West Coast of the United States. Even if we had not abandoned Pearl Harbor, a Japanese air and naval base at Midway would have denied the free use of our major naval base in the Pacific to our forces.

Attack or Wait?

Without question, the Battle of Midway turned on Nagumo's decision to rearm his Kates and Vals. Even so, upon learning of the presence of an enemy carrier, Admiral Yamaguchi on board the *Hiryu* urged Nagumo to attack immediately with dive-bombers.

However, without a proper fighter escort to accompany the Vals, Nagumo elected to delay his attack until he could send out a larger, more balanced group.

Q: *What if Nagumo had kept his second wave armed with torpedoes? Was it possible for Nagumo to deliver a limited strike of Vals, as advised by Yamaguchi?*

Jon Parshall

Here are a couple of typical Midway "what if" scenarios that illustrate the main problem with most of their ilk; namely, that they're just *way* too simplistic.

Any Midway counterfactual scenario that attempts to ferret out Nagumo's options in the 6:30 to 9:30 a.m. time frame *must* posit the following:

1. When are the Americans detected, and when does that information reach Nagumo?
2. Where/how far away are the Americans, and are the Japanese in a position to attack them?
3. Has Nagumo given/not given/given-but-retracted his re-arming order for an additional attack against Midway?
4. What is the status of 1st Carrier Division's torpedo planes? Are all/some/none of them armed with torpedoes?
5. What is the location/status of Lieutenant Joichi Tomonaga's Midway strike force? When will it be returning for landing operations?
6. What American attacks are transpiring at the time that Nagumo receives his initial scouting reports and makes his decision to counterattack?
7. What is the status of the Japanese combat air patrol? How many aircraft are aloft? Has an attack transpired recently that would require the force to recover and re-arm fighter aircraft and launch new CAP replacements?

Without an understanding of these seven variables at the very least (and there are probably more), one cannot make a meaningful exploration of *any* given counterfactual scenario about Midway, period. *All* of these variables are interrelated, and *all* of them have a direct bearing on if, when, and how the Japanese could have launched a counterstrike.

The other fundamental problem that many Midway counterfactual story-spinners gloss over is that they assume that the antiship strike aircraft were just sitting up on Nagumo's flight decks, ready to go at a moment's notice. According to this view, all Nagumo would have had to do was wave his white-gloved hand and a hundred aircraft would magically be winging their way toward the Americans. Nothing could be further from the truth. In fact, the reserve aircraft were all in the hangars throughout the morning of June 4. This meant that from the time Nagumo said "Go," it would have taken a minimum of thirty to forty minutes (forty-five or so being the norm) before those aircraft could have been lifted to the deck, spotted, warmed up, armed (in some cases), and launched, during which time the flight decks would have been completely shut down to any other air activities. This is why understanding what else is going on in the battle with those other variables is crucial to understanding what options Nagumo really had available to him at any given time. Without defining those variables, no meaningful discussion is possible on the topic. Of course, a counterfactual scenario this complex is way over the head of most armchair strategists (it's probably over my head, too, frankly), which is why so many Midway "what ifs" are just useless rubbish and a complete waste of pixels.

Incidentally, this helps illustrate just what a horrible pickle Nagumo was really in. He, too, had to evaluate each of these variables. Not only that, but he had to weigh them on the basis of imperfect information (namely an incomplete scouting report from *Tone*'s fourth scout plane that gave an incorrect location of the enemy force) while standing in the tiny, noisy fishbowl of *Akagi*'s bridge, surrounded not only by his own staff, but also by the ship's company (whom he would not have wanted to lose face in front of by asking too many questions of his staff that might reveal his lack of deftness as a carrier admiral). He had just gotten through one series of American air attacks, and within seven minutes of his receiving his scouting report he would be under attack again by incoming American B-17s. So he had about ten minutes, under the worst possible conditions, to make a clear, reasoned determination as to what to do. Is it any wonder that he got it wrong?

Frank Snyder

If Nagumo *had* kept his reserved torpedo planes armed with torpedoes, he would have been able to launch torpedo planes immediately upon learning of the presence of American ships (including one American aircraft carrier—probably *Enterprise*, not *Yorktown*). But if he *had* launched a strike of bombers and torpedo planes, they would have been unescorted, and therefore vulnerable to any defending fighters that the Americans had wisely retained for defense against air attack.

When the Japanese scout plane sighted the American carrier (probably *Enterprise*, inasmuch as that ship reported sighting a Japanese scout plane at about the same time), it probably should have seen *Hornet* as well, but it would not have seen *Yorktown*, which was still trying to catch up with the other two carriers. I doubt that Nagumo would have acted differently if he had been told that there were *two* American carriers in the vicinity. The lack of fighters to escort an immediate strike would probably have deterred him.

John Lundstrom

Nagumo simply didn't believe that the Americans would show up that quickly at Midway. I also believe that when the Americans appeared, the Japanese really weren't overly concerned. In my opinion, I think the Japanese simply didn't view the Americans as much of a threat. Nagumo didn't think the Americans would be at Midway in great numbers. The situation was very much analogous to the situation Spruance had at Philippine Sea. At that battle in 1944, Spruance decided to let the U.S. carrier force absorb the Japanese strike and, after surviving it, he then launched his own attack on the Japanese. I believe that was Nagumo's philosophy at Midway. Nagumo also expected Army bombers to be based at Midway, along with patrol planes and fighters. He didn't expect carrier dive-bombers to be on the island. So when the Marine dive-bombers hit the *Hiryu* a little after eight o'clock on the morning of June 4, that's when Admiral Yamaguchi realized that an American carrier was nearby. Eventually, he recommended that the Japanese send off an attack on the American carrier before landing the returning Midway strike force. The *Hiryu* and *Soryu* probably could have sent a strike force off their decks before their Midway planes were in bad shape as far as gasoline left in their tanks. They could have sent the dive-bombers off alone. I think Nagumo saw no reason to take that step. He made a conscious decision to absorb any American attacks and then he would launch his own attack and destroy the American ships.

Ronald W. Russell

It didn't happen because the Japanese mind-set prohibited attacking ships with Kates that were not armed with torpedoes, and to a slightly lesser extent, with Vals not armed with armor-piercing bombs. But had Nagumo launched at least the Vals with their land bombs, accompanied by a Zero escort, it's highly likely that some important damage could have resulted on one or more of the U.S. carriers. This is very evident from the success a very few Vals and Zeros had on the afternoon of June 4 (a direct hit with three bombs on *Yorktown*). The land bombs wouldn't have been as damaging as armor-piercing ordnance, but they would have caused *some* damage and would have disrupted U.S. fleet operations.

A Full Strike

Despite the two-hour scouting advantage that the Americans possessed, could Nagumo have emerged from Midway a big winner? If he had not rearmed his torpedo planes with bombs, could Nagumo have delivered the knockout blow to the U.S. Pacific Fleet?

Q: *If Nagumo had been able to launch his grand assault on the American carriers the morning of June 4, what kind of damage would the Japanese have inflicted?*

Jon Parshall

This is completely unknowable, of course, and depends on a variety of factors, so there's no way one can really issue a blanket "answer" of some sort. The American carriers, of course, are operating in two separate groups. Will the Japanese find both groups? Or just one? If just one, which one? How effective will the American CAP be against a multi-altitude attack? What's the weather like? How well is American radar working? Is damage control on board a newer, less experienced ship like *Hornet* going to be as good as that on board the *Yorktown*? There are a zillion different variables.

Notwithstanding the preceding, there's no question that at this point in the war, nobody did combined-arms aerial attacks like the Japanese navy. If they had managed to catch the Americans in a group, my sense is that they would have attacked two of the carriers, most likely with a squadron apiece of dive-bombers and torpedo aircraft. If so, there's every possibility that the Americans would have lost both those ships outright. You'll note that when two American carriers were attacked by two deck-loads of Japanese carrier aircraft at Coral Sea, we lost one carrier (*Lexington*) and almost lost the second (*Yorktown*). Double the number of Japanese aircraft committed (four deck-loads), and the likely outcome is two American carriers sunk.

Peter C. Smith

Yorktown would have been sunk without a doubt, and maybe Spruance's pair would have been discovered, hit, and either heavily damaged or made inoperable, if not sunk outright. That would have made Yamamoto's cruiser and battleship groups totally secure against meaningful air attacks and they could have done whatever they wished.

Ronald W. Russell

This comes up repeatedly. One needs to remember that Japanese expectations were that no U.S. carriers would be in the Midway vicinity—why would they be? Absent the U.S. radio intelligence that the Japanese never knew about, they would not have been. The Japanese expected to march into Midway à la Wake Island, rouse the U.S. carriers that were assumed to be in port at Pearl, and defeat them when they rushed north to take back Midway. That was their only focus. The presence of Task Forces 16 and 17 on June 4 was totally outside their box, something which they had neither war-gamed nor expected. That being the case, they had no plans to launch a fully coordinated strike against carriers when the battle commenced.

Summary

Yamamoto was deeply committed to the Midway operation, to the point that he refused to wait for *Shokaku* and *Zuikaku*. And while Coral Sea temporarily cost the Japanese two additional fleet carriers, American repair crews worked feverishly to make *Yorktown* serviceable. So instead of potentially enjoying a 2-to-1 advantage in carriers at Midway, Nagumo outnumbered the United States just 4-to-3 in large carriers.

Nagumo was also operating in the dark with multiple objectives. With no clear-cut idea as to where enemy ships might be located, Nagumo had to subdue Midway's defenses, avoid the island's attack planes, and put himself in position to overwhelm the American carriers, if they made an appearance.

The Americans had shown up unexpectedly at Coral Sea. You have to wonder why the Japanese would not safeguard themselves from a repeat occurrence just a month later by thoroughly scouting the waters around Midway and by keeping aircraft armed and ready to deal with enemy carriers.

Whether it came down to overconfidence, a lack of respect for the enemy, or a rigid adherence to timetables, the failure by the Japanese to prepare for a worst-case scenario would be their undoing.

Seven

RISING SUN ECLIPSED: MIDWAY II

AT 6:30 A.M. ON JUNE 4, more than one hundred Japanese bombers pounded Midway. In response, Midway's attack planes went after Nagumo in waves and then the U.S. carriers joined the fight, as Spruance launched sixty-eight dive-bombers, twenty-nine torpedo planes, and twenty fighters from *Hornet* and *Enterprise*.[1]

The next few hours would be crucial. With the Americans bearing down on him, Nagumo prepared to deliver a fully coordinated strike. All he needed was time.

A Horrific Slaughter

Midway attempted to deliver the first blow. Six Avengers, four B-26s armed with torpedoes, two sections of Dauntless and Vindicator dive-bombers, and fifteen B-17s attacked one after another; all failed to score a hit.

Up next were the three Devastator torpedo squadrons from *Hornet*, *Enterprise*, and *Yorktown*. *Hornet*'s Torpedo 8, led by Lieutenant Commander John Waldron, attacked without fighter escort at 9:30 a.m., followed minutes later by *Enterprise*'s Torpedo 6, commanded by Lieutenant Commander Eugene Lindsey. Then, around 10:15 a.m., *Yorktown*'s Torpedo 3, with Lieutenant Commander Lance "Lem" Massey in charge, made its run with limited fighter protection. In eight separate attacks, five by Midway's planes and three by carrier-borne torpedo squadrons, not one bomb or torpedo hit a Japanese ship.[2]

Midway's planes suffered grievous losses, as did the slow-moving Devastators from *Enterprise*, *Hornet*, and *Yorktown*. Only four of forty-one Devastators survived; all fifteen of *Hornet*'s torpedo planes cartwheeled into the sea, with pilot George Gay the lone survivor. Some might have called the courageous assault suicidal, but in his book, *Sole Survivor*, Gay claimed that it would have been foolish for the torpedo planes to flee the scene without attacking.[3]

Despite the absence of hits, the determined attacks kept Nagumo off-balance and drew the Japanese combat air patrol to lower altitudes.[4] Much like a boxing match, the Americans had flailed away at their opponent with numerous body shots. Although individually ineffective, collectively the attacks placed dive-bombers from *Enterprise* and *Yorktown* in position to deliver the knockout blow.

Q: *What if the Devastators from* Hornet, Enterprise, *and* Yorktown, *and Midway's attack planes, had not pressed home their attacks?*

Ronald W. Russell:
"What if" speculation on the Battle of Midway is an ever-popular exercise for two fundamental reasons. First, the battle was an absolutely stunning event in the conduct of the Pacific War. In the space of only a few hours on a single day during the war's sixth month, the mighty Imperial Japanese Navy was abruptly and unexpectedly thrust from offense to defense for the remainder of the conflict. And second, that only happened because of a sequence of very remarkable circumstances that dictated the battle's unlikely outcome. Had any one of those circumstances not occurred exactly as it did, the result would surely have been far worse for the American side.

Among those remarkable circumstances, seven stand out as essential in order for the United States to prevail at Midway:

1. American code-breakers *had* to be able to read the IJN's fleet radio code, an incredibly complex, multilayered cipher system that the Japanese believed throughout the war to be unbreakable.
2. The Japanese carriers *Shokaku* and *Zuikaku had* to sustain enough damage at the Battle of the Coral Sea to keep them out of Admiral Yamamoto's Midway operation (his original intent was to include them, bringing six fleet carriers to Midway instead of four).
3. The shipyard crew at Pearl Harbor *had* to get the USS *Yorktown,* mauled in the Coral Sea, ready for action in the three days mandated

by Nimitz instead of the three months they had originally estimated as necessary.

4. Torpedo Squadron 8 (VT-8) *had* to break away from the wayward attack course of the USS *Hornet*'s air group, thus becoming the first U.S. fleet squadron to find the enemy carriers and setting off a chain of events that eventually destroyed them all.
5. The Japanese destroyer *Arashi had* to be drawn far from the enemy carrier force in pursuit of submarine USS *Nautilus,* thus becoming an unwitting and critically needed guidepost for USS *Enterprise* dive-bombers.
6. The Japanese combat air patrol (CAP) *had* to be lured to the wave-tops by successive attacks from American torpedo bombers, thus leaving the upper altitudes undefended when the dive-bombers arrived in force.
7. Finally, the dive-bombers from both the *Enterprise* and the *Yorktown* *had* to arrive over the enemy carrier force at the same time.

If *any* of the above seven circumstances had not occurred as they did and when they did, then the likelihood is very high that instead of four Japanese carriers destroyed on June 4, 1942, the total would have been three, two, one, or even none. And regardless how the Battle of Midway would have proceeded in any of those scenarios, the end result could only have been worse for the United States and its allies—possibly much worse, even to the point of losing the battle entirely and surrendering the atoll to the invaders.

Among the above seven key drivers of America's victory at Midway, one stands out for its implausibility, the sole event among them that absolutely should not have happened: VT-8 skipper Lieutenant Commander John Waldron's blatant defiance of his immediate superior while engaged in the midst of combat during a war. Waldron was convinced at the outset that the course to the Japanese fleet ordered by his air group leader, Commander Stanhope Ring, was critically flawed. Upon launching from the *Hornet,* Waldron followed Ring for a time, then abruptly changed his squadron's heading, vowing to find the enemy carriers if he had to do it all alone. As it happened, his intuition was exactly right. VT-8 scored a near bull's-eye in tracking the intended target, although all of its planes were lost with no damage to the enemy. But the large clouds of smoke generated by the Japanese ships during their fight with Waldron were sighted by the *Enterprise* torpedo squadron, VT-6, then on a tangential course that very nearly caused them to miss the Japanese themselves. Spotting the smoke from VT-8's immolation, VT-6 altered course in order to engage, but suffered essentially the same fate as their counterparts from the *Hornet*.

However, that fight generated still more smoke from the Japanese ships, which in turn caught the attention of a torpedo plane gunner from the *Yorktown*. His squadron, VT-3, was similarly on a course that otherwise would have bypassed the Japanese carriers. The gunner called his pilot's attention to a faint wisp of dark smoke on the horizon, the pilot relayed the sighting to his squadron commander, and VT-3 turned to engage. Their fate was similar to the previous two VTs, but there was a crucial difference in the case of the *Yorktown* airmen: their dive-bomber squadron, VB-3, was high overhead and tracking VT-3's heading. When the torpedo planes altered course, VB-3 followed suit and thus was able to find their eventual target, the Japanese carrier *Soryu*, which they then destroyed. They did so successfully, without fighter opposition, because the Japanese CAP was then occupied with VT-3 at low altitude. And at the same moment, VB-6 and VS-6 (Scouting Squadron 6) from the *Enterprise* were accomplishing the same thing vis-à-vis the carriers *Akagi* and *Kaga*. (Alone and shorn of many of its planes, the remaining Japanese carrier *Hiryu* suffered a similar fate later that day.)

That is how the Battle of Midway actually played out and why it was a smashing victory for the U.S. side. Knowing those facts, we can now make reasonable postulations as to what very likely would have happened in the absence of Waldron's disobedience. Had he followed his commander's orders (eminently sensible in the midst of combat!), the battle on the morning of June 4, 1942, would have proceeded in a vastly different manner. Commander Ring had led his air group far from the U.S. task force, beyond the fuel range of his fighter escort, all of which had to ditch. The torpedo bombers of VT-8, as short-legged as the fighters, would also have been lost in the sea. Ring's dive-bombers barely made it back to the *Hornet* (some diverted to Midway) with a few drops of gas in their tanks.

But without the annoyance of VT-8's attack, the Japanese carrier force and its escorting cruisers and destroyers would not have had to slow their advance to fend off a torpedo assault, and most importantly, the escorts would not have thrown off those billowing clouds of signaling and maneuvering smoke that were spotted by VT-6. If the *Enterprise* squadron had thus continued on its tangential course well to the southeast of the Japanese, their flight that morning would have been as inconsequential as that of Commander Ring. And that, of course, would have denied VT-3's gunner his wisp of dark smoke on the horizon, leaving the *Yorktown* squadrons no reason to deviate from their own southerly course. That would have obviated VB-3's attack on the *Soryu,* virtually assuring its survival that morning, if not throughout the battle.

By no means does the bad news stop there. The repetitive torpedo attacks by VT-8, VT-6, and VT-3 had drawn successive flights of the Japanese CAP down to low altitude and kept them there while the American dive-bombers formed up overhead unscathed. Without the torpedo men's sacrifice at the wave-tops, the defending fighters would have had no reason to abandon their usual stacking over the fleet. That would have resulted in the two *Enterprise* dive-bomber squadrons being met with a formidable defense that surely would have improved the survivability of *Akagi* and *Kaga,* and indirectly the *Soryu.* It is very difficult to imagine two squadrons (thirty-one planes) being able to neutralize all three of the Japanese carriers in the face of a properly positioned and alert CAP. And remember, VB-3 is not a factor in this scenario—they are following VT-3 on its errant course to the southeast, leaving *Soryu* to fight on.

The result, then, of Waldron simply following orders, as any reasonable officer in like circumstances would have done (and in fact, all but Waldron actually did), would certainly have been the survival of one or more of the three carriers sunk on the morning of June 4. And the survival of even one of them would have greatly enhanced the chances for *Hiryu* to withstand any follow-on attacks later in the day. By sundown on the fourth, the Japanese would have been left with at least two functioning carriers, perhaps even three or four. In the actual battle, the *Yorktown* was mortally wounded from a single desperate attack by only a handful of shot-up Japanese dive-bombers and torpedo planes. How would the *Enterprise* and *Hornet* have fared on the following day if, for example, *Soryu* and *Akagi* had been able to field three or four times more attackers in the air than *Hiryu* had sent against *Yorktown?* Nothing positive comes to mind in that event.

With the singular exception of Waldron's defiance of Ring, all of the individual circumstances leading to the "Miracle at Midway" were within the realm of reasonable expectations: the code-breakers could and did achieve exceptional success, *Shokaku* and *Zuikaku* could have been and in fact were sidelined as the result of Coral Sea, the yard workers at Pearl Harbor did manage to get *Yorktown* ready for sea through exceptional around-the-clock effort, *Arashi* could and did inadvertently help the *Enterprise* dive-bombers find their way, and all three dive-bomber squadrons could and coincidentally did arrive over the Japanese fleet at the same time. But Waldron's act was the anomaly of them all, the clear exception to the rule, the happening that should not have happened. Yet it did nonetheless, and the course of the battle, the war in the Pacific, and much of history to follow was thereby profoundly changed.

Frank Snyder

Anyone who believes that torpedo planes are going to orbit around and wait for someone else to show up in the presence of many Japanese Zeros close to their bases does not understand the fundamentals of survival in the air in 1942. If the *Hornet* torpedo squadron had waited for the *Hornet* dive-bombers, it would have waited forever—or until Zeros shot them down—because the *Hornet* dive-bombers had gone in (what turned out to be) the wrong direction.

The *Enterprise* torpedo squadron (flying at a lower altitude) apparently could see the smoke of the Japanese ships, and had "cut the corner" to arrive at the Japanese while *Enterprise* bombers were conducting an "expanding box" search for the Japanese. If the *Enterprise* torpedo squadron had waited for the dive-bombers before conducting an attack, the outcome would have depended on the altitude at which the Japanese fighters were located. If the Japanese fighters were already at a lower altitude (to have dealt with the *Hornet* torpedo squadron, for example), the losses in the *Enterprise* torpedo squadron might have been greater. If the Japanese fighters had been waiting at a higher altitude to intercept American dive-bombers, the losses in the *Enterprise* torpedo squadron might have been less, but bombing by American dive-bombers might have been less effective, while the chances of effective torpedo attacks might have been better.

The attack by *Yorktown*'s torpedo squadron was more nearly a coordinated attack than were attacks by the other torpedo squadrons, but the comments above—about how the altitude of the Japanese defensive fighter planes would have affected the losses and effectiveness of the *Enterprise* torpedo squadron—also apply here.

Lew Hopkins

I was in Bombing 6 flying off the *Enterprise* and we had absolutely no fighter opposition when we dove our planes on the *Kaga* and the *Akagi* the morning of June 4 until after we had made the attack. If the three American torpedo squadrons had not pressed home their attacks against the Japanese carriers earlier in the morning and drawn the Japanese fighters down to sea level, it would have been a different story. If the Japanese fighters had been at a higher altitude to challenge us, I doubt whether the dive-bombers could have knocked out the three carriers that morning. The Japanese had ample fighters protecting their carriers but they lacked central control. Each carrier had its own combat air patrol up, and all of the fighters converged on our torpedo planes and failed to protect their carriers from the dive-bombers.

Sumner Whitten

If the early-morning June 4 attacks against the Japanese carriers had not been aggressively undertaken, then more Japanese fighters would have been available for the defense of the Japanese carriers against the dive-bombers. As it was, just the VMSB-241 (Marine dive-bombing squadron) rear gunners accounted for seven to eight Japanese fighters, to say nothing of disrupting the re-arming and recovery of Japanese fighters and bombers returning from the Japanese attack on Midway. The Japanese would have had more fighters in the air and their carriers would have been ready for our dive-bomber attack later in the morning.

Dusty Kleiss

If Midway's bombers and the American torpedo squadrons had not pressed home their attacks, the odds in our favor would have been diminished. Our best chance of slaughter was while the Japanese carriers were launching and recovering planes, particularly planes that were running out of fuel. The Japanese required more time to refuel their aircraft and to re-arm them than the Americans did. The Japanese knew that our B-17s, with their Norden bomb sights, could easily sink their slow-moving transports. Our Marines, with obsolete battle-damaged planes, gave great help in disorganizing the four main Japanese carriers. Without such help, our success odds would have been fifty–fifty.

Lieutenant Commander John C. Waldron, standing third from left, was the commanding officer of Torpedo Squadron 8 (VT-8) on USS Hornet.

John Waldron

Of the three U.S. carrier-based torpedo squadrons, *Hornet*'s Torpedo 8 usually gets most of the attention, perhaps because all fifteen Devastators were lost with just one survivor.

Maybe it's because of VT-8's skipper, Lieutenant Commander John Waldron, who exhorted his men to drive home their attacks, even if they were the last ones left alive in the squadron.[5]

In making his ill-fated charge, Waldron disregarded a direct order from the commander of *Hornet*'s air group, Commander Stanhope Ring, and broke away from *Hornet*'s dive-bombers and fighters to seek out the enemy.[6] Ring found nothing and returned safely; Waldron, who would posthumously receive the Navy Cross, found the Japanese and lasting fame.[7]

Q: *What if John Waldron had survived the Battle of Midway?*

Barrett Tillman

None of the three American torpedo squadron commanders survived Midway: Waldron of VT-8, Massey of VT-3, or Lindsey of VT-6. But the latter two died following orders from *Yorktown* and *Enterprise* air groups. For reasons bound to remain unknown, John Waldron broke off from *Hornet* air group's formation and led his fifteen Devastators to the Japanese carriers, perishing to everlasting glory. As noted elsewhere, the air group commander, Stanhope C. Ring, failed to find Kido Butai and squeaked back to TF-16 nearly out of fuel. A few of his ship-borne dive-bombers (SBDs) and all of his escorting F4Fs splashed with dry tanks.

The extent of *Hornet*'s institutional failure at Midway was a well-kept secret for many years, partly because of the late-war reputation of the skipper, Marc Mitscher. But the fact remains that only *Hornet*'s VT-8 found the enemy on June 4, whereas the other carrier groups plus Army and Marine units on Midway also attacked the target.

If in fact Waldron disobeyed orders and pursued his own instincts, how might he have been regarded if by some miracle he survived the battle? "The Court Martial of John C. Waldron" could make for compelling drama.

Stanhope Ring was well known as a stiff-necked disciplinarian: "martinet" is not too strong a description. Before the battle, he grounded two experienced SBD pilots who remained seated because they did not notice him enter the wardroom. Reportedly, he owed his position to a prewar friendship with Mitscher, and in fact had been slated to depart *Hornet* because Waldron was next senior. (How the air group might have performed under a more competent and aggressive commander makes for intriguing speculation.) Assuming that Torpedo 8's skipper had survived after acting independently, contrary to his air group commander's wishes, what then?

In reality, it is almost inconceivable that the Navy would have been stupid enough to charge Waldron, a genuine hero who had attacked the most important target afloat, especially when the commander of the air group (CAG) ran airplanes out of gas with no damage to the enemy. The negative public image for the service could have been devastating. However, recall that the Navy court-martialed Captain Charles McVay of the sunk cruiser *Indianapolis* in 1945, even bringing the Japanese submarine commander as a prosecution witness. McVay remained the only captain so treated, though 350 other U.S. Navy ships were sunk during the war. The usual conclusion is that Fleet Admiral Ernest King prosecuted the unfortunate cruiser skipper in order to cover up institutional failings throughout the service.

The outcome of a Waldron trial would have depended wholly upon John Waldron, not the U.S. Navy. As a devoted professional, he might have chosen to accept the role of scapegoat and sacrifice his career "for the good of the service." Perhaps he would have bided his time, awaiting the postwar era to clear his name.

On the other hand, if he had chosen to fight in court, he would have looked unbeatable with an objective jury.

Any defense Waldron mounted would have turned on the workings of *Hornet*'s leadership, especially Mitscher and Ring. Waldron's attorney could have asked some extremely uncomfortable questions:

Why did Mitscher launch his short-ranged fighters first, thus limiting their range?

Why did fighter skipper Samuel Mitchell run ten Wildcats out of fuel, with two pilots drowned?

Why did Ring continue his original course, ignoring Waldron's accurate navigation?

Why did the ship's action report contain demonstrable falsehoods?

Depending upon the timing of the trial, Waldron's acquittal could have had long-term effects for naval aviation. With proof of Mitscher's lax leadership style and Ring's culpability, would either of them have been promoted again? Ring's subsequent career was unremarkable other than the fact that he retired with three stars. But Mitscher's status as commander of the fast carriers in 1944–45 could have been scuttled in a 1942 or 1943 court battle, with unknowable consequences. Certainly, the U.S. Navy would have triumphed in the Pacific regardless of who led Task Force 38/58, but the success achieved under Mitscher was significant. (In truth, he relied more heavily upon his staff than was generally realized for half a century.)

More likely, John Waldron would have received early selection for captain and commanded a carrier in 1944–45. Whether he retired as a flag officer or not, his place in history would only have grown in the years following the Battle of Midway.

Peter C. Smith

Two alternative scenarios seem to apply here. 1) He may well have been court-martialed for a) disobeying his commanding officer totally, b) breaking the strict radio silence while in the air on the way to the target against all known standing orders, and c) sacrificing his entire command without result. Or, 2) he may have gotten the Purple Heart and promotion for outstanding gallantry in the face of the enemy. It would have depended on what made the best headlines back home, and a brave, heroic sacrifice was right for the time after six months without any major U.S. victory. History has treated him very well; hard-facts analysis is still very much resented and notably not faced up to.

Lost at Sea

While Torpedo 8 found the Japanese, the rest of *Hornet*'s attack group, led by Commander Stanhope Ring, had no such luck.

Initial sighting reports mentioned only two Japanese carriers. Perhaps Ring was searching for the other two; in the end he drew a blank.[8] Ring returned to his carrier with nineteen other Dauntlesses, while thirteen SBDs landed at Midway, and three dive-bombers and all ten of Fighting 8's Wildcats ditched at sea.[9]

Dive-bombers from *Enterprise* and *Yorktown* eventually blasted three of the Japanese fleet carriers, but Ray Spruance would later claim that if *Hornet*'s SBDs had discovered and successfully attacked the fourth carrier, *Hiryu*, *Yorktown* might have survived the battle.[10]

Q: *What if* Hornet*'s entire air group had hit Nagumo in a coordinated attack?*

Ronald W. Russell

It has often been speculated that had the two dive-bomber squadrons from USS *Hornet* managed to attack the Japanese carriers, the battle would have turned out even better for the U.S. side. Presumably, the *Hiryu* would have been neutralized in addition to the other three carriers that were attacked during the morning battle on June 4. However, that notion assumes an unlikely level of good luck and bombing skill on the part of the *Hornet* airmen, and it discounts a critical timing factor that actually would have severely played against their compatriots from the other carriers.

For VB-8 and VS-8 to find the Japanese fleet, they would have needed to follow John Waldron and VT-8 to the target. And therein lies the timing problem. Recall that VT-8 was the first American carrier squadron to appear over the enemy carriers, at about 9:25 a.m. (Midway local time) on the morning of June 4. If the *Hornet* dive-bombers had accompanied them to the target at their cruising altitude of about 18,000 feet, they would have been met by the Japanese CAP at more or less the same altitude. The critical problem there is that, in the actual battle, there were *no* U.S. carrier planes menacing the Japanese fleet at high altitude until the *Enterprise* and *Yorktown* SBDs arrived nearly an hour later. Consequently, the Japanese fighter pilots were conditioned to the need for reacting to a repetitive threat from low-flying torpedo planes—first VT-8, then VT-6, and finally VT-3.

But that would not have occurred if dive-bombers from the *Hornet* had appeared at high altitude at the outset. Yes, the Japanese CAP would have had to deal with a multi-dimensional attack, but whether the *Hornet* squadrons themselves would have significantly threatened the enemy carriers is by no means assured. The U.S. Mark 13 torpedo was notoriously unreliable against maneuvering ships (the *Shoho* had been sunk by torpedoes in the Coral Sea only after dive-bombers had left it dead in the water), and neither VB-8 nor VS-8 had ever practiced bombing a moving target. So even with the Japanese fighters forced to contend with VF-8 plus dive and torpedo bombers at the same time, the likelihood of significant damage to their ships in this scenario is rather low. In all likelihood, the four Kido Butai carriers would have survived the attack, possibly with some damage, but still capable of continuing operations.

If we then assume that the *Enterprise* and *Yorktown* squadrons arrived on scene when they actually did, they would have faced a Japanese CAP that had already been alerted to the threat of carrier-borne dive-bombers at high altitude, thanks to the *Hornet* air group an hour before. The threat posed by VT-6 and VT-3 would still have drawn many of them to the wave-tops, but it's reasonable to believe that at least some of them, thanks to their experience with the *Hornet* attack, would have maintained some level of high-altitude stacking to await the inevitable dive-bombers that seemed to follow the TBDs (torpedo bombers). Any amount of Japanese CAP meeting the *Enterprise* SBDs at their altitude could very likely have thwarted Richard Best's incredibly lucky, off-the-cuff assault on the *Akagi* (in the actual battle, he managed to take the ship out of the war with only two bombs from three of his planes).

Thus, we are left with a supreme irony: in the actual event, Waldron's unaccompanied attack set off a chain of events that led to the destruction of three Japanese carriers as the battle had barely commenced. But if the *Hornet's* other squadrons had joined him in the same attack, the enemy quite possibly would have fared much better before the day was over, with the ultimate outcome left very much in doubt.

This unique set of circumstances illustrates a point that I frequently make whenever discussing alternative history scenarios for the Battle of Midway: virtually anything you can contrive other than what actually happened would in one way or another have turned out better for the Japanese.

Frank Snyder

In order for the *Hornet*'s dive-bombers to have been "fortunate" enough to find the Japanese, Mitscher (having been selected for flag rank, and about to be relieved as the commanding officer of the *Hornet*) would have had to send off *Hornet*'s dive-bombers to the southwest instead of to the west, but that did not happen.

At the time that strike planes in both *Enterprise* and *Hornet* were launched, neither Spruance (in *Enterprise*) nor Mitscher (in *Hornet*) knew where the Japanese carriers were, nor in what direction the Japanese had steamed since they had been sighted about an hour earlier. And they did not realize that in fact the Japanese ships had been forty miles farther away from Midway than initially reported.

Fletcher had actually directed Spruance to attack the Japanese "when definitely located," but Spruance apparently felt that the initial position would be sufficient, and assumed that the Japanese would continue to steer toward Midway at roughly full speed, so he ordered *Enterprise* squadrons on a course southwest to intercept the Japanese. Mitscher, on the other hand, while also assuming the initial reported position to be correct, reasoned that any later movement by the Japanese was unpredictable, so he ordered the *Hornet* group to proceed westward toward the Japanese position reported earlier, and should the Japanese not be there, to head southeast toward Midway.

Spruance's staff (actually, it was Halsey's staff, temporarily supporting Spruance), apparently unused to overseeing the operation of *Hornet*'s air group, was focused primarily on the *Enterprise* air group, so that the *Hornet* air group was allowed to seek out the Japanese on their own.

As it turned out, Spruance had guessed correctly, and Mitscher had not. In the event, the Japanese *did* continue toward Midway, although they were behind their estimated position for several reasons: their initial position as reported by an American seaplane put the Japanese forty miles closer to Midway than they actually were; the Japanese speed had been reduced during the time that the Japanese were maneuvering to avoid the American planes from Midway; and once the Japanese planes returning from their strike on Midway had landed, Nagumo had changed the course of his Japanese formation from southeast to northeast.

Following *Arashi*

As *Hornet*'s Dauntlesses flew their mission to nowhere, *Enterprise*'s dive-bombing squadrons, Bombing 6 and Scouting 6, led by the commander of the *Enterprise* air group, Lieutenant Commander Wade McClusky, were also having a tough time tracking down the Japanese. Arriving at the interception point and finding nothing but empty ocean, McClusky carried out a box search.

Finally, at 9:55 a.m., low on fuel, McClusky spotted an enemy destroyer, *Arashi*, which had just chased off the U.S. submarine *Nautilus* and was sailing northeast at high speed to rejoin Nagumo's carriers.[11]

McClusky tagged along and hit the jackpot, sighting the enemy fleet, and at 10:25 a.m. the SBDs began their steep dives. Doctrine called for Lieutenant Dick Best's Bombing 6 to attack the closest flattop and Lieutenant Earl Gallaher's Scouting 6 to dive on the next carrier in line. Instead, McClusky, Scouting 6, and several of Bombing 6's SBDs pounced on the nearest carrier, *Kaga*, forcing Best to switch targets and attack *Akagi* with just three Dauntlesses.[12]

In a matter of minutes, *Enterprise*'s dive-bombers devastated the two carriers. Earl Gallaher's bomb was the first to slam into *Kaga*, followed by at least three others, including one by Lieutenant (junior grade) Dusty Kleiss. Dick Best's 1,000-pound bomb nailed *Akagi* while two others were either a hit or a near-miss. Aircraft on *Akagi*'s hangar deck erupted in flames, setting off a fatal chain reaction of explosions.

Both carriers were doomed; *Kaga* sank that evening, *Akagi* slid under the waves the following day.[13]

Q: *If McClusky had not spotted* Arashi, *would* Enterprise*'s dive-bombers have discovered the Japanese carriers on their way back to their carrier? What if Dick Best had not attacked* Akagi?

Lieutenant Commander Clarence Wade McClusky, Enterprise*'s air group commander, who led his Dauntlesses to* Akagi *and* Kaga.

Frank Snyder

When McClusky spotted *Arashi*, he was just about to make a turn to the right—generally to the northeast—but the sighting of *Arashi* allowed McClusky to adjust his new course so that he would be heading directly toward the Japanese force.

I believe that even without sighting *Arashi*, McClusky would have flown close enough to the Japanese force to have spotted them, although his squadrons might have attacked the Japanese ships slightly later than the attack by the *Yorktown* dive-bombers—who had arrived from a nearly opposite direction.

Lew Hopkins

If our air group commander, Wade McClusky, had not decided to follow *Arashi*, we probably would have returned to *Enterprise* without ever having dropped our bombs, if we had enough gas to get back to the carrier. Most of the bombers we lost off *Enterprise* on the morning of June 4 ran out of gas. If the *Yorktown* sinks the *Soryu*, then it's three against three.

Let's face it. The Japanese were better trained at that point in time than the Americans were. I was an ensign and had been out of flight training for nine months. I had not flown the SBD, our dive-bomber, until March 1942. I was one of a number of ensigns involved in the Battle of Midway and it turned out that it was the seasoned veterans who did most of the damage against the Japanese carriers at Midway. One of those seasoned veterans was Dick Best, the skipper of Bombing 6, who hit *Akagi* with his bomb. Most of our squadron, Bombing 6, attacked *Kaga* but Dick Best, fortunately, successfully attacked *Akagi*. I'm not really sure how it happened that most of Bombing 6 attacked *Kaga*. I was flying in the second division and I just followed the division leader. I didn't hear any radio communications that supposedly took place between Best and McClusky when it came to the assigning of targets and I don't think many people heard it.

Dusty Kleiss

First, let us assume that Lieutenant Commander McClusky saw *Arashi*. Then we must also assume that Earl Gallaher and Dick Best had also seen her going at full speed. Those two had keen eyesight and were super-observant of everything. Now let us assume that McClusky checked his YE-ZB (tracking/navigation system), detected the direct bearing to the Big E, and went directly in that direction. I am 99.5 percent certain that the only ones that would have followed him would have been his two wingmen. Dick and Earl would have done an even better job of torching *Kaga* and *Akagi*. There would have been no confusion as to who should dive what.

Had we never seen *Arashi* and been forced to head for home empty-handed, we might easily have lost a couple of our carriers and many crews.

Had Dick Best not broken off his attack on *Kaga*, and diverted to *Akagi*, that would have been very bad for us. *Akagi*'s planes, in addition to *Hiryu*'s planes, would have attacked the American carriers. There is also one other possibility. Our first few bombs on the *Kaga* hid her from view, due to the intense fire and smoke. Very likely, some of our planes would have automatically shifted over their attack to *Akagi*. This had happened in other situations.

Now consider what would have happened if Admiral Halsey had been present. Earl Gallaher would have led the pack; we would not have circled and wasted lots of gas. We would have known that it was totally unlikely that the Japanese carriers would speed up; they were likely to have slowed down. Had we arrived at the expected maximum location, we would have definitely assumed that they had slowed down. Had we encountered bad weather, we would have made a mile-wide search pattern using signals without radio, as we had practiced many times. McClusky had never practiced this and probably didn't know this procedure.

Why do I say Earl Gallaher would have led the pack? Halsey put men into their positions based on past performances rather than on seniority. Let me give a couple of examples. Sometimes we received aviation mechanics that were inept or lazy. We would immediately promote them upward to a rating where we were in excess. Then we were "forced" to send them to some shore station having a vacancy. They had time and facilities; we didn't. Here's another example: I frequently flew the group commander's plane on scouting hops, although several other pilots were senior to me. It was important to keep that plane fully tested, and to make sure that all things were operational. Why me? I had managed to find our carrier with a faulty compass and a radio that didn't work. I managed to find our carrier, despite the fact that the weather had become so bad that a new "point option" was sent out by radio to a new moving recovery point. Also, I had managed to make a good carrier landing with an SBD that was considerably shot up with many bullet holes.

Halsey would have selected McClusky to head our F4F fighter group where he really knew his forte. He would have done a better job than Jim Gray. He also would have done a masterful job taking down *Hiryu*'s planes. He previously had won a Distinguished Flying Cross for his skill as a fighter pilot in the Marshall Islands.

Barrett Tillman

History often turns upon small coincidences. The unexplained death of Genghis Khan's son at Vienna in 1242; the fog that allowed Washington's defeated army to escape Manhattan in 1776; if Hitler had been among the millions killed in World War I.

Then there was His Imperial Japanese Majesty's ship *Arashi*.

Commissioned in October 1940, *Arashi* was a *Kagero*-class destroyer. She had fought in the Java Sea battle of February 1942 and was assigned to the Midway operation. On the morning of June 4, she was detached from the carrier force to pursue the contact of the American submarine *Nautilus*. Unsuccessful in the hunt, she was returning to Kido Butai when her wake drew the attention of Lieutenant Commander Wade McClusky, commanding *Enterprise* Air Group. At his back were some thirty SBDs.

McClusky had departed *Enterprise* at 7:45 a.m., flying 150 miles southwest toward the expected contact point with the Japanese carriers. Arriving at the briefed position, he found nothing and therefore initiated a "box search." He reckoned he could continue the hunt until 10:00 a.m. before dwindling fuel would force him to return.

At 9:55 a.m., the Big E's CAG saw *Arashi*'s wake, proceeding northeast. McClusky figured she was a cruiser on a liaison mission between the occupation force and the carriers, and turned to follow her course. In a matter of minutes, he was looking at Nagumo's striking force.

Absent the *Arashi* sighting, it is speculation to say whether McClusky would have continued his search beyond 10:00 a.m. But since at least one-sixth of his Dauntlesses splashed from fuel exhaustion, it is prudent to assume that he would have returned to the ship rather than press his search.

In that case, the battle would have gone in Japan's favor.

Without Bombing 6 and Scouting 6 to destroy *Akagi* and *Kaga* that morning, the absolute best that can be assumed for the Americans is that somehow Bombing 3 still got to *Soryu*. In that case, the afternoon matchup would have pitted *Akagi* and *Kaga*, against Task Force 16 and 17, and *Hiryu* probably would have escaped destruction at least long enough to participate in a coordinated attack. Considering what *Hiryu* alone did to *Yorktown*, the odds of *Enterprise* or *Hornet* surviving *Akagi* and *Kaga*'s big air groups appear thin.

Perhaps Japanese scouting and communications would have complicated Nagumo's problems even more than they actually did. Perhaps there would have been enough time for Task Force 16 to retrieve *Enterprise* and *Hornet* aircraft and launch a strike before the enemy squadrons arrived. In that case, the carrier battle conceivably could have been a draw. But without three intact flight decks, the Pacific Fleet likely could not have supported offensive operations that summer.

Having made her brief appearance on history's stage, *Arashi* exited the spotlight. Ultimately, she was sunk in the Battle of Vella Gulf in August 1943.

Q: *If Devastators from* Hornet, Enterprise, *and* Yorktown *had not kept the Japanese combat air patrol occupied between 9:20 and 10:20 a.m., how would an alert Japanese CAP have dealt with Dauntlesses from* Enterprise *and* Yorktown?

Anthony P. Tully

Asked in this way, the role of the torpedo plane attacks in wearing down and distracting the Japanese CAP is hard to overstate, especially the role of VT-6 (*Enterprise*'s) and VT-3 (*Yorktown*'s) torpedo squadrons. If those attacks had not taken place, or if all the torpedo squadrons had attacked before 10:00 a.m., there would have been considerable danger that the Japanese CAP would have regrouped and succeeded in intercepting one or more of the approaching dive-bomber squadrons and inflicting immediate early losses. However, as the experience with *Hiryu* in the afternoon and other times in the war would show, it was difficult for the Japanese to fully interdict or prevent the attack of a skilled and determined dive-bomber force. In such a scenario, it seems likely that *Kaga* would still have been hit and only slightly less likely *Soryu* too, though both carriers might have received fewer hits. In their condition, with hangars jammed with armed and fueled aircraft, fewer hits might not have mattered, or saved them. Yet by the same token it seems a very safe bet that *Akagi* would not have been hit at all. So if the torpedo planes had not distracted the Japanese CAP, at the very least two carriers (*Akagi* and *Hiryu*), and possibly three (if *Soryu* had been missed), would have been left to launch the 11:00 a.m. strike against TF 17 and TF 16. In such a scenario, the intangible factor of Nagumo's flagship remaining fully operational, and the consequent effect on Japanese morale and momentum, calls into serious question just what would have happened with both *Akagi* and *Hiryu* slugging it out with *Enterprise*, *Hornet*, and *Yorktown* the rest of the day. Further, this scenario is a very conservative one, simply assuming that the removal of the torpedo bombers' sacrifice results in *Akagi* alone surviving. It is entirely possible that only one—or none—of the four Japanese carriers would have received critical damage if the CAP had been in place.

Peter C. Smith

In my opinion, even at full strength, the Japanese CAP, good though it was, would not have been able to totally prevent at least a portion of damage. *Some* American dive-bombers would have gotten through and scored hits, but probably not so many. The real key here is bomb hits would not normally have sunk these Japanese carriers on their own; it was the fact that the ships were packed belowdecks with fueled-up aircraft fully laden with bombs, plus additional bombs and torpedoes lying around and not correctly stowed away, which stoked the furnaces when hits were taken. *That* was why the three carriers were lost as a result of the first dive-bomber attacks. Bear in mind that one of their carriers was hit several times at the Battle of Coral Sea, but managed to get home safely to Japan afterwards. It usually took a combination of bomb hits and torpedo strikes to put these ships down.

Jon Parshall

If the U.S. carrier torpedo plane attacks had not occurred when they occurred, the Japanese most likely would have been able to launch a counterstrike against the U.S. fleet before the U.S. dive-bombers arrived. That would have exposed the U.S. carriers to a much greater level of danger in the 11:00 a.m. time frame. It also would have removed large numbers of fueled and armed aircraft from the Japanese hangars, thus improving their passive damage control in the event of a bombing attack. That said, even with the Japanese CAP at optimal altitude, with three squadrons of U.S. dive-bombers attacking across two vectors, the odds are that at least one Japanese carrier would have been hit. That's just complete speculation, though—it's really completely unknowable.

Four for Four

While *Enterprise*'s SBDs took care of *Kaga* and *Akagi*, Dauntlesses from *Yorktown* pounded *Soryu*. Several hits turned the carrier into a flaming wreck; she sank that evening.

One carrier, *Hiryu*, temporarily escaped destruction, perhaps assisted by Fletcher's decision to hold back *Yorktown*'s scout-bombers. Midway's Catalinas initially reported only two enemy carriers and Fletcher kept his scouting squadron in reserve, just in case additional carriers were discovered.[14]

Q: *If* Yorktown*'s scout-bombers had been sent in the morning attack, would all four Japanese carriers have been hit?*

Frank Snyder

If *Yorktown*'s scouting squadron *had* been sent with the bombing squadron *and* both these squadrons had arrived over the Japanese at the same time that both *Enterprise* squadrons arrived, then it is possible that all four Japanese carriers would have been destroyed, but it is also possible that *Hiryu*, the fourth carrier, would have been far enough away from the other carriers to avoid being detected and attacked.

Earlier, the Japanese aircraft carriers had been maneuvering—often independently—to defend themselves, first from air attacks by planes from Midway and then from attacks by the three torpedo squadrons that attacked separately. The Japanese were not using radar for early warning or tracking, so for the Japanese, defense against air attacks was decentralized.

Peter C. Smith

A good question. Indeed, instead of a futile (and very short-range) search to the north by a portion of *Yorktown*'s fighting strength (a task that could have been much more effectively conducted by the accompanying heavy-cruisers' scout planes or the Midway-based PBYs, types of aircraft which would not have revealed the presence of American carriers to the Japanese had they been spotted themselves), the whole strength of *Yorktown*'s air group could have been sent away and may have nailed *Hiryu*, thus preventing her two counterattacks that did in the *Yorktown*.

Command Decisions

With *Akagi*, *Kaga*, and *Soryu* out of action, it was now up to *Hiryu* to carry on the fight, and her planes did just that, bombing and torpedoing the *Yorktown*. Though *Yorktown* was initially abandoned, boarding parties later tried to save the American carrier, but torpedoes from a Japanese sub eventually sent her to the bottom.

SBDs from *Enterprise* and *Yorktown*, flying from *Enterprise*, gained vengeance that same afternoon, slamming four bombs into *Hiryu*, which sank on June 5.[15]

Even without his fleet carriers, Yamamoto still possessed powerful surface ships that could devastate the Americans in a night battle. Yamamoto hoped to lure his enemy westward, but Spruance, who had replaced the ailing Halsey, refused to take the bait. To protect his carriers and Midway, Spruance moved eastward on the evening of June 4, away from Yamamoto's trap.[16]

After two more days of bombing fleeing warships and sinking a cruiser, *Mikuma*, Spruance cashed in his chips and retired to the east, bringing down the curtain on the Battle of Midway.[17]

Q: *What if Halsey had been in command at Midway, rather than Spruance?*

Barrett Tillman

William F. Halsey was not known as "The Bull" for nothing. Loud, aggressive, and flamboyant, he proved himself more a fighter than a thinker. But in the post–Pearl Harbor period, when fighting admirals were thin on the water, Chester Nimitz valued Halsey's willingness to come to grips with the enemy.

Beached with severe dermatitis, Halsey was unable to deploy to Midway. But fortunately for the American cause, he recommended Rear Admiral Raymond Spruance, his screen commander, to take his place, and Nimitz concurred.

Certainly Halsey performed well during the early hit-and-run phase of the Pacific War, as Task Force 16 struck far-flung Japanese garrisons in the Marshalls and Gilberts. The Big E also had escorted *Hornet* during her "delivery" of the Doolittle Raiders in April, but those were small, sometimes inconclusive actions with nothing remotely comparable to the stakes at Midway.

Halsey handled all the tasks handed to him throughout 1942, but as a fleet commander from 1944 onward he was far less capable. He oversaw three debacles in just a few months: the misstep at Leyte Gulf in October, "Halsey's Hurricane" in December, and another storm in January. The common factors in each were the admiral's extreme aggressiveness and inattention to critical factors, especially intelligence. For reasons still unclear, he ignored information on the powerful Japanese surface force transiting San Bernardino Strait toward Leyte Gulf, and dismissed meteorologists' concerns about the two storms, with fatal consequences.

Nevertheless, opinion is divided as to how Halsey might have performed at Midway. The main concern is usually focused on the situation beginning

Rear Admiral Raymond Spruance, who avoided Yamamoto's trap and sailed east.

the evening of June 4, when the Japanese carriers had been destroyed. Based on his later behavior, the Bull would have wanted to pursue the beaten enemy and inflict even greater damage, as he did at Leyte. Such action was devoutly wished by Admirals Nagumo and Yamamoto, who hoped to draw the Americans into the kind of battle to the Imperial Japanese Navy's liking—a night surface engagement.

In reality, Spruance decided against an all-out pursuit, drawing bitter criticism from his air staff but high marks from Nimitz, who had stressed two points: defend Midway and operate according to "the principle of calculated risk." Ray Spruance played his hand to perfection. What would Halsey have done?

More importantly, what would Nimitz have done? Would he have given Halsey the same latitude that Spruance received, or would he have read his subordinate properly and kept a guiding hand on the conn? Probably not the latter, considering that Nimitz retained Halsey not only after Leyte but also after both storms in 1944–45. In fact, Halsey was so well established that he even received a postwar promotion to (five-star) fleet admiral despite the heavy losses he incurred (five ships with 720 men at Leyte; three destroyers and eight hundred men in the first typhoon).

Another factor is the advice that Halsey received from his staff. After all, Halsey was not a "real" aviator like Marc Mitscher, Jocko Clark, and Arthur Radford: he had received his wings only in 1935 when there were not enough flying captains to meet increased requirements. Consequently, he relied heavily upon his subordinates. His mercurial chief of staff, Captain Miles Browning, clashed with the aviators at Midway and (properly) lost a dispute with Spruance. It is far from certain that Halsey would have overruled his longtime chief of staff regarding critical matters such as mission timing and tactical radius.

So let us assume the worst: Halsey's air groups destroy four carriers on June 4 and run toward Yamamoto's battleships two days later. Would American scout planes have spotted the threat in time to avert disaster? Probably—if Yamamoto had hauled within range during daylight. But at night—that's another matter. The battle could have ended a tie, with serious problems for American plans later that year.

In any case, America can remain grateful that in June 1942, Frank Jack Fletcher and Raymond A. Spruance had the conn in Task Forces 16 and 17.

Hal Friedman

I think if Vice Admiral William Halsey had been in command at the Battle of Midway, as opposed to Rear Admiral Raymond Spruance, the battle might have ended in a defeat, and possibly a disaster, for the United States. I obviously cannot prove this counterfactual scenario, but Halsey's record of aggressiveness is well known and it might have succeeded in snatching defeat from the jaws of victory.

Part of the evidence comes from later battles in which Halsey was in command. In the Battle of the Santa Cruz Islands in October 1942, Halsey had the *Enterprise* and *Hornet* attack a superior Japanese carrier force that resulted in *Hornet* being sunk and *Enterprise* nearly so. These losses can be justified as part of the defense of Guadalcanal, but they also demonstrate Halsey's aggressiveness even in the face of enemy superiority. The even more obvious case was the Battle of Leyte Gulf where Halsey squandered very significant American naval superiority to destroy the Japanese carriers while leaving large portions of the Japanese surface fleet to threaten U.S. amphibious forces. Halsey, in spite of these failings, was part of the Brown Shoe Club that criticized Black Shoes such as Spruance, Fletcher, and Admiral Thomas C. Kinkaid for not understanding carrier air power and being too skittish in its use during the war.

Imagine what might have happened at Midway if Halsey had been in command by the end of the major action on June 4, 1942, when *Yorktown* was disabled and would soon be sunk, along with the destroyer *Hammann*, by the Japanese submarine I-168. The remnants of the American task forces numbered less than two dozen warships, and on the evening of June 4 even fewer than that since numerous destroyers were still attending crippled *Yorktown*. Imagine further if vintage Halsey had come out that evening and sailed west in pursuit of the remnant of the Nagumo force. Given the absolute inferiority of American surface night-fighting capabilities at this time in the war, the obvious Japanese superiority in that venue, and the fact that Nagumo was spoiling for a night surface battle, Halsey might have sailed into Nagumo's two battleships, three cruisers, and remaining destroyers with the highly vulnerable *Enterprise* and *Hornet* nicely silhouetted against the night sky. It's not clear if Nagumo could have destroyed the American force since he would have been outnumbered in cruisers and destroyers, but his escort ships probably would have given Halsey's escorts quite a run for the money, while the two Japanese battleships could have easily dispatched the two remaining American carriers. In the morning, this would have reduced American airpower to numerous PBYs and B-17s (both not very useful against naval surface forces) and about a dozen and a half Wildcats, Dauntlesses, and Vindicators on Midway. By this time, Nagumo would have been reinforced by the Kondo force and the next day by the Yamamoto force. Though this would have still been primarily a surface force, it would have been extraordinarily powerful. Such a force would have been more than capable of bombarding Midway and landing forces, not to mention cleaning up remaining American naval surface units. Such a force would also have been able, with light carrier forces of its own that could have been reinforced in a couple of days by the light carriers of the Hosogaya force from the Aleutians, to truly dominate the region. All in all, it might have proven a disaster for the United States, with the loss of its remaining carriers and Midway itself.

Spruance was also criticized for his conduct at the Battle of the Philippine Sea when he allegedly demonstrated timidity in going after the Combined Fleet, just as at Midway, as the criticism went. That criticism should have ceased after the Battle of Leyte Gulf, but much of this was Brown Shoe criticism of the Black Shoes without thinking about what might have been. The near-disaster at Leyte Gulf might have been a certain one at Saipan, but at least in those two battles the United States enjoyed general superiority. The opposite was the case at Midway. Spruance made the prudent (not timid) and correct decision to turn east, keep Midway covered, and keep his remaining carriers and escorts intact. This, I think, is the most unfortunate and even tragic aspect of picking our naval heroes by their personalities and public images rather than their operational accomplishments. Spruance, the "Quiet Warrior," made the right play at the right time, did not publicize it, and largely remained silent when criticized for it. There was no bluster, no headlines, and no fanfare, just a naval professional doing the right thing for his country when it needed him most. How many Americans today know about "Bull" Halsey or might recognize his picture if shown it? Then, how many will know about Raymond Spruance and understand what he and many other quiet men did for the country? Raymond Spruance reminds me every day when I teach history about the people in the past who did things for us even though they did not know us—and the number of people today enjoying security and prosperity who have no idea what was done for them by these quiet people from the past.

Frank Snyder

If Halsey *had* been present at Midway, he would have commanded the American force from his flagship *Enterprise*. Spruance would have been present in a cruiser, and Fletcher might or might not have been at the battle.

Halsey was more aggressive, so it is not clear whether he would have flown the air search to the northeast, as ordered by Nimitz and carried out by Fletcher. What is also in doubt is what Halsey would have done when he received the initial report of the sighting of the Japanese carriers—from a seaplane. Would he have held all three carriers to await the return of scouts before heading to intercept the Japanese? Or would he have departed with all three carriers immediately, trusting that returning scouts could find him? Or would he—like Fletcher—have detached two carriers to steam off to strike the Japanese, and have retained one carrier to land the scout planes?

It seems unlikely that he would have—as Spruance did—proceed in the evening *away* from the Japanese, and return to be off Midway on the following morning. It is more likely that, at the end of daylight on June 4, he would have pursued the remnants of Nagumo's force, and risked an encounter with Kondo's battleships, also closing on the remnants of Nagumo's force, a movement of which Halsey would probably have been unaware. He might or might not have been spared a night action with Japanese surface forces—which had to change course once Yamamoto issued his order for all Japanese (except the submarine U-168) to withdraw.

Invading Midway

If Nagumo had somehow won the carrier battle, or if Japanese transports had attempted to invade Midway despite the IJN's disastrous loss of carriers, how would the invasion have fared?

Midway's defenses included the 6th Marine Defense Battalion, with additional reinforcements from Companies C and D of the 2d Raider Battalion. Ground forces could also count on the Marine 3d Defense Battalion's 37-mm antiaircraft battery and plenty of guns, ammo, dynamite, mines, torpedo boats, and barbed wire.[18]

In December 1941, it had taken the Japanese nearly two weeks to overcome approximately five hundred Americans on Wake Island. Would Midway have been any different?

Q: *What if the Japanese had attempted to invade Midway?*

Frank Snyder

If, despite a loss of their aircraft carriers, the Japanese *had* carried out their plan to invade Midway, they might have succeeded. Resistance by the Marines would have been quite strong, and the U.S. PT boats deployed at Midway might have taken quite a toll on Japanese transports. The air groups in the two operational American carriers should have been decisive, but I doubt that air groups focused on the destruction of enemy aircraft carriers and other naval ships would have contributed much to forestall an amphibious landing or to support ground action ashore.

The great success on June 4 cannot be assumed for air attacks against ships other than aircraft carriers. Aircraft carriers tend to sink themselves, once their own fuel starts to burn and their bombs start to explode. On June 4, the Japanese aircraft carriers were hit by an average of three bombs apiece, but two days later, on June 6, thirteen American bombs plus a plane crash were required to sink a Japanese cruiser.

If the Japanese had succeeded in occupying one or both of the islands at Midway, the course of the war would have changed markedly. The U.S. submarine effort would have been refocused on the interdiction of the Japanese LOC to Midway, and the United States might have been reluctant to deploy any aircraft carriers very far from Hawaii. It is possible that Midway rather than Guadalcanal would have become the "whirlpool" into which both sides would pour more and more naval forces.

John Gardner

If the American code-breakers had not been successful in breaking the Japanese code, and the U.S. fleet had not steamed to the north of Midway in position to ambush the Japanese fleet, Midway's small force of Marines, with no naval vessels, would not have been able to defend the island for more than a couple of days. They would have been attacked by five aircraft carriers carrying 259 aircraft, accompanied by two hundred ships, including battleships, cruisers, destroyers, and submarines, and a five-thousand-man landing force. It must be remembered that Midway's air group consisted of a squadron of obsolete, ineffective Brewster Buffalo fighter aircraft and a squadron of old obsolete SBD2U-3 dive-bombers. When it was determined that Midway was the target, Admiral Nimitz reinforced Midway with what was available to him, a number of mostly obsolete aircraft and about two thousand Marine infantrymen.

Prior to the intelligence breakthrough that caused actions to reinforce Midway, the U.S. Marine Corps' 6th Defense Battalion, consisting of fewer than six hundred men, would have been hard-pressed to hold the island for any length of time, but morale was high and we would have inflicted many casualties against a landing force. Marines are, first and foremost, infantrymen and trained to fire rifles and machine guns. The IJN, attacking with even one aircraft carrier and supporting cruisers and destroyers, would have been able to lie just outside the range of our 5-inch guns and would have hammered us round the clock until we would have had no artillery. We would have been forced to rely on small-arms fire and bayonets, waist deep in water. Those were

Colonel Harold Douglas Shannon's orders: "There will be no surrender. We will fight them waist deep in the water with bayonets."

The Japanese plan was to kill, to destroy, and to occupy Midway. That accomplished, they would have been able to patrol all of the north Pacific to Alaska, and south to within four hundred miles of Oahu. Following the defeat of Japan in 1945, it was learned that their battle plan was to regroup and assemble at Truk, resume operations early in July, and capture New Caledonia and the Fiji islands, as well as Guadalcanal in the Solomons. Building an airstrip at Guadalcanal would have given the IJN a base that would have enabled their aircraft to choke the sea-lanes between the U.S. West Coast, New Zealand, and Australia. In August, they had planned to attack Johnston Island, Palmyra, as well as Hawaii, with the possibility of occupying the big island of Hawaii and building an airstrip, which would have been capable of supporting attacks on all of the Hawaiian Islands.

At this point, Pearl Harbor and Oahu would have been severely threatened. Supplying necessities to Hawaii would have been a nightmare. The civilian population and the military would have been clamoring for food and supplies. People in need of food would surely have demanded President Roosevelt to sit down to a bargaining table and negotiate with the Japanese.

The president's plans to support an all-out attack and defeat of Hitler would have been put on hold, while the war in the Pacific would have received more troops and supplies. The war in Europe to defeat Hitler would have been slowed. The delay would have given Hitler more time to develop an atomic bomb. Had Hitler developed the atomic bomb, everyone knows he would have used it. The British Islands would have been devastated. Any Allied offensive would have been shelved. President Roosevelt would have gone back to another negotiations table.

The U.S. Navy's victory at Midway was a victory that saved the United States of America. I find it futile to find an argument that could logically dispute that the Battle of Midway was among the most important in the history of the United States of America.

Jon Parshall

In my opinion, the Japanese would have been bloodily repulsed by the American defenders. The Japanese landing force was undersized for its mission, and outnumbered by the American defenders. Because of the nature of the atoll, the Japanese landing barges would have had to deposit the landing forces on the reef, several hundred yards offshore, whereupon the Japanese infantry would have had to wade through chest-high water (or deeper) all the way to the beach. The beaches themselves were very well defended with obstacles, barbed wire, and a plethora of homemade mines. The Marine defenders were well dug in and copiously supplied with both heavy and automatic weapons. They also had a platoon of light tanks in reserve on Sand Island. The landing forces, on the other hand, were essentially light infantry, and were equipped with no organic firepower heavier than light mortars and a few medium machine guns. The Japanese navy had no doctrine for either aircraft or gunfire support for landing troops, meaning that they wouldn't have been able to give good support to the landing forces, even if those forces made it alive to the beach. The Japanese had no reserve troops or contingency plans should the initial landings go badly. In sum, any dispassionate analysis of the operation can't help but lead to the conclusion that the Japanese would have been slaughtered wholesale to little purpose.

Sumner Whitten

Even though the Japanese had lost four carriers, and had a couple of destroyers and cruisers damaged, they still could have attacked Midway. We were anticipating such action and were making plans to become another Wake because we really didn't know how badly the Japanese carriers had been hit. But I believe the Japanese still could have taken Midway, even though we still had some fighters and bombers left, and even though the U.S. Navy still had a couple of carriers available for action and still packed a big punch, without any air opposition from the Japanese. The Japanese had battleships and big cruisers for land bombardment, as well as some smaller vessels. Of course, our submarines would have been able to take a shot at those ships too. But the Japanese could have taken Midway if they had *really* wanted to do so.

Barrett Tillman

Dawn, June 5, 1942. *Yorktown* and *Hornet* are sunk. *Enterprise* is limping homeward, battered by bombs and torpedoes and pursued by Japanese submarines. The Marines on Midway are standing by to repel invading Japanese forces.

Midway was defended by 1,700 Marines, plus 1,300 Navy and Army support personnel. The Japanese landing force's five thousand troops therefore lacked the traditional 3-to-1 numerical advantage commonly required for an attack against prepared defenses. However, an examination of the composition of the landing force causes even more concern. The force involved army and navy units that had never trained together—there was no practice landing—and who lacked detailed knowledge of the atoll's defenses. The dedicated combatants may have numbered only 2,500 or so, with a similar number assigned support roles. Even discounting the U.S. Army and Navy personnel, 1,700 entrenched Marines versus 2,500 attackers yields an offense-to-defense ratio of less than 1.50. That is half the 3-to-1 figure advocated for most offensives, and often considered marginal in amphibious operations.

There is more to the equation than the initial success of a Japanese amphibious victory. Assuming that the attackers could have seized Midway, the Japanese would have found themselves in the position of the proverbial dog that chases the automobile down the block. Even if he catches the car, now what?

In the weeks and months after the battle, the Imperial Navy undoubtedly could have developed the atoll into a useful advanced base. But conducting offensive operations against Hawaii would have been a huge task. Midway lay one thousand nautical miles from the nearest Japanese base at tiny Wake Island, 1,400 miles from the Marshalls, and 2,200 miles from Japan. The logistics required to sustain a worthwhile garrison on Midway would have been substantial.

Almost immediately, the Pacific Fleet would have moved against Midway. A submarine blockade—even with the marginal torpedoes available in 1942—would have complicated Tokyo's effort to keep the new conquest adequately supplied. And from early 1943 B-17Fs with "Tokyo tanks" were available to make the ten-hour round-trips from Hawaii. Thus, a Japanese garrison on Midway would have been subjected to frequent attacks, probably including nocturnal bombardment by battleships and cruisers. In contrast, the Imperial Navy could not have defended its advanced base with its own warships, as Midway lacked room for docking large vessels.

Bottom line: Japanese possession at Midway probably would have proven more a liability than an asset, and Midway could not be held indefinitely.

Rising Sun Victorious

Midway turned out to be a signature victory for the United States and a crushing defeat for Japan. With the loss of four fleet carriers, a heavy cruiser, more than three hundred planes, and 3,500 men, the Japanese had suffered a shocking setback. While the Americans lost the *Yorktown*, a destroyer, 150 planes, and 307 men, the United States had also wrested the initiative away from Japan.[19]

What if Yamamoto's plans had unfolded without a hitch and Nagumo had crushed the Americans at Midway? How could Nimitz possibly have hoped to contest Japanese operations in the latter half of 1942?

Q: *If Japan had won at Midway, can you envision future Japanese operations and how Nimitz might have responded?*

Jon Parshall

Fiji and New Caledonia were already on the docket. They would have gone for those first. Port Moresby likely would have been next, as this would have essentially sealed off the approaches to Australia. After that, Hawaii would have been next up, although clearly that was never within the realm of possibility. There was no way that was going to happen—the Japanese didn't have the necessary ships, logistics, amphibious doctrine, or sufficient sealift to get enough troops to the point of attack and land them on a hostile beach.

With one or two carriers remaining for Nimitz, the goal would have reverted to protecting Hawaii and perhaps contesting Japanese operations aimed at the remaining outposts of Hawaii, like Johnston Island.

Peter C. Smith

Although talk of the Japanese occupying Hawaii was widespread and popular, that occupation was not, in my opinion, logically feasible over such a distance. There were insufficient ground troops available, not enough ships to carry them, the Japanese army did not have the will to commit them, and they could not have been sustained or supplied. Fiji, New Caledonia, and Samoa would have been the logical progression, cutting Australia off from U.S. aid. These gains could have been fully supported by a chain of airfields along the Solomons.

With a limited amount of fleet carriers, Nimitz would have been forced to hunker down at Pearl and the West Coast ports. The Americans would have had to withdraw many more ships from the Atlantic to appease U.S. civilian clamor to protect the West Coast and to concentrate on trying to keep Australia supplied, if possible. This would have been coupled with submarine attacks on the extended Japanese lines of communications.

With the strong exception of Admiral Ernest King, the American chiefs of staff were totally fixated on and determined to carry out the "Second Front" landings in France as early as 1942 or 1943, and they fought very hard to have this policy adopted under the "Germany First" agreement. It took an awful lot of argument by their British opposite numbers to persuade them, most reluctantly, that such a policy was premature due to Allied military (as distinct from naval and air) weaknesses in men and matériel, and that it made more sense to adopt the North African and Sicilian landings instead while building up their power. A defeat at Midway may have reinforced the stance of American chiefs of staff in this regard, although public opinion at home, especially west of the Missouri, would probably have been just the opposite.

John Lundstrom

If the Japanese had captured Port Moresby and Midway, then what would the Americans have done? It might have depended on the reaction of the American public. The attack on Pearl Harbor turned the public's attention away from the war in Europe. Would the United States simply have shut down operations in the Pacific? Eisenhower, when he was in the War Plans Division early in 1942, envisioned simply shutting down operations in the Pacific, essentially holding a certain line in the South Pacific from Australia to Hawaii, and then going to defeat the Germans first. It makes you wonder if a defeat at Midway would have resulted in the United States totally focusing on Germany or perhaps, if the public demanded it, an increase in focus on defeating the Japanese. The Americans weren't going to get additional aircraft carriers until 1943. The American philosophy was to save resources for the war in Europe until 1944 when the United States had enough resources for both Europe and the Pacific.

It is highly debatable whether the Japanese could ever have taken Hawaii had they captured Midway. They were never going to be able to interdict the supply line between Hawaii and the West Coast. The Japanese were never strong enough to take Hawaii. They certainly got to the point where they thought they were strong enough and were making plans to capture Hawaii.

Sumner Whitten

If the U.S. Navy had not possessed intelligence of an impending Japanese attack, no American carriers would have been nearby and the Midway garrison and aviation units would not have been on the alert. Therefore, the Japanese would have encountered no opposition to an attack on Midway.

The Japanese attack on Midway would have come as a complete surprise, both to Midway and to Pearl Harbor. The Japanese would have reduced Midway's defenses, made a landing with but very little effective opposition, and suffered only a minor loss of men and materials.

This scenario would have made Pearl Harbor vulnerable, somewhat, from long-range bombers, of which the Japanese had but few, which would in turn have made Pearl Harbor and all of Hawaii a large base, totally armed, and with the American carriers in good shape. It also would have mandated that all American carriers in the Atlantic be transferred to the Pacific, thus reducing U.S. aid to Britain.

The Japanese would have been able to continue to advance throughout the Pacific and then attack, at least by air, New Zealand and Australia. This would have affected the British in Egypt and in the Mediterranean. It would have also stopped or reduced American aid to Europe for at least a year. What would have been the effect on Britain?

Perhaps to send a warning, Japanese submarines would have attacked and harassed Mexico and Latin America. Would this have drawn those countries into the war more deeply or would it have made them more reluctant to aid the United States?

Instead of Guadalcanal in 1942, we would have been back to Midway in 1943–44, just to relieve Pearl Harbor of attack. We would have retaken Midway, considering our manufacturing capabilities, mileage, submarines, and bombers capable of hitting Midway. Perhaps, after retaking Midway, we would have gone on to retake Wake Island and then into the Philippines, rather than all of the other island campaigns which originated from our Australian/ Guadalcanal bases.

Europe would have had to hang on for at least nine to eighteen months until the United States rebuilt its fleets and air power and expanded its Army.

After retaking Midway, the Americans would have continued to launch offensive operations, but on a reduced basis until 1944. The war would have lasted until 1947 or until Russia became dominant.

Barrett Tillman
Assuming a complete Japanese victory over the Pacific Fleet, America would have been left with one large carrier, *Saratoga*, while *Wasp* would have been transferred from the Atlantic. That was insufficient to support Operation Watchtower, the Guadalcanal landings in August, and in fact would have required the United States to put the entire Pacific War on hold. No offensive action would have been possible until the latter part of 1943, when wartime construction began arriving in the form of *Essex*-class and *Independence*-class carriers. In that regard, whether Japan seized Midway was almost irrelevant. America's most valuable asset was the mobile striking power of the fast carriers, far more than two islands more than a thousand miles from Hawaii.

Looking downstream, assuming that the historic schedule would have been maintained beginning in the fall of 1943, USS *Missouri* would not have dropped anchor in Tokyo Bay until the summer of 1946. During the interim, Tokyo could have consolidated its gains, conducted operations elsewhere (perhaps against Australia), and fortified both the outer and inner ring of island garrisons. Amphibious operations, always potentially sanguinary, would undoubtedly have been even bloodier. Again assuming a similar chronology, the Marianas would not have been secured until mid-1945, shortly before the atomic bombs became available. In that regard, Little Boy and Fat Man would have become world-historic wild cards, and surely America would have used more nuclear weapons as they were produced.

Whatever the delay in defeating Japan, the war in Europe would have continued. But the Allied "Germany First" strategy undoubtedly would have been re-examined. With America reeling in the Pacific, and no immediate way to redress the situation, it is not difficult to envision immense political pressure to defend Hawaii and the West Coast. In 1942, rare assets destined for North Africa and the Mediterranean almost certainly would have been diverted westward rather than across the Atlantic, thus affecting the progress of the war against Germany and Italy. That situation would have benefited the Soviet Union.

The geopolitical map of Europe would likely have been redrawn in the wake of a Japanese victory at Midway. It is not certain that the Normandy landings would have been possible in 1944, perhaps leading to greater Anglo-American emphasis on Italy. Meanwhile, the Russian steamroller would have continued bulldozing its way westward, paying an even higher price in blood and treasure against a Wehrmacht less concerned with defending the Atlantic Wall. By the time of VJ Day in 1946, the Soviet T-34s might well have parked in France's northern department of Pas de Calais, and the postwar world that we knew would have been rendered almost unrecognizable. The implications for the Cold War are many, varied, and unknowable.

A somewhat different situation would apply had Admiral Nimitz decided not to defend Midway with his carriers. With Midway in Japanese hands (assuming the island was seized), U.S. naval operations in the Hawaiian area would have been affected but not prevented. Guadalcanal still could have been occupied, and the historic Pacific timetable largely maintained.

Summary

Midway ranks as one of the most historic naval battles of all time. In just five minutes, American dive-bombers knocked out three Japanese carriers and dramatically reversed the direction of the Pacific War.

Several factors created the perfect opening for *Enterprise*'s and *Yorktown*'s Dauntlesses, including determined American torpedo attacks, McClusky spotting *Arashi*, Nagumo's rearming order, and sloppy handling of ordnance by the Japanese. If Nagumo's CAP had been at a higher altitude, if *Enterprise*'s Dauntlesses had missed *Arashi*, if Nagumo had not rearmed, or if the IJN had limited damage to its carriers, perhaps June 4 would have been Nagumo's day.

Instead, Midway changed the complexion of the war; within two months, the Americans would launch their own offensive in the southwest Pacific.

Eight

ISLAND OF DEATH: GUADALCANAL

THANKS TO MIDWAY, the Americans were ready to begin the long, bloody trek to Japan, starting with Guadalcanal and Tulagi in the Solomons. Both were attractive targets, the former because of its new airfield, and the latter due to its excellent natural harbor. To protect the supply line with Australia, and to safeguard Allied bases to the east, the United States was prepared to launch its first amphibious assault of the Pacific War.[1]

A Matter of Timing

After occupying Guadalcanal in June 1942, the Japanese immediately began to construct an airfield that would allow their bombers to reach Allied bases in New Caledonia, the New Hebrides, Fiji, and possibly Samoa. The Americans were not about to let that happen.

A U.S. task force of more than eighty ships, including several aircraft carriers, pounded Guadalcanal on August 6–7. Then 19,000 Marines of the 1st Marine Division, led by General Alexander Vandegrift, assaulted Guadalcanal and Tulagi. Tulagi fell in a matter of days; Guadalcanal would take much longer.[2]

Initially, the Marines met little resistance in capturing the Japanese airfield, later christened Henderson Field. However, over the next six months, Japan would attack by air, land, and water in a desperate effort to eject the invaders. Two carrier battles, Eastern Solomons and Santa Cruz, and several surface actions, including a major Japanese victory off Savo Island on August 9, left the Americans hanging on by a thread. On the ground, the Marines repulsed several enemy assaults on August 21 and again in September and October.

In early February 1943, the Japanese called it quits, evacuating some 13,000 sick and starving survivors from what some called the Island of Death, irrevocably reversing the tide of the Pacific War.[3]

Q: *What if the Americans had not captured Guadalcanal or had delayed their invasion? Could Japan have made the island impregnable and eventually attacked Fiji or Samoa?*

John Lundstrom

If the United States had delayed its offensive on Guadalcanal, it would have been much more difficult to carry out, had the Americans allowed the Japanese to consolidate their positions on the island.

As the Americans were putting together their plans for the offensive, they received intelligence that the Japanese were fortifying their positions on the island at a fast pace. If the Americans had delayed their offensive perhaps by just a few weeks, they might have had to fight their way into the island because at that point the Japanese might have had a functioning air group on Guadalcanal. The Americans would have lost more carriers quickly in the battle for the island and that might have shut down U.S. air operations. The brilliance of going into the Solomons was that once you have a toehold on the island chain, your land-based air would be within range of a number of other objectives. You wouldn't need carrier planes as much at that point.

As far as the Japanese were concerned, half their carriers weren't ready to go when the battle for Guadalcanal started. Their second carrier division wasn't ready to come to Guadalcanal. They were still training aviators. You also have to wonder about the fuel constraints of the Japanese capital ships.

I don't believe Yamamoto felt Guadalcanal was going to be as decisive a campaign as it turned out to be. He really hesitated to commit the full strength of his battle fleet. He never fully grasped the importance of his capital ships shelling the U.S.-held airstrip on Guadalcanal. Of course, Nimitz really didn't want to send his battleships to Guadalcanal either. He was hesitant to send battleships into the shallow waters around the Solomons.

Gordon Rottman

I do not think that the Japanese would have been able to retain Guadalcanal. Even if the United States had suffered an even more significant naval defeat than Savo Island or the Marines had experienced serious reverses ashore, the Japanese did not have the forces or logistics available to exploit any successes.

If the United States had delayed the invasion for even a couple of months, it is doubtful the Japanese would have made significant progress in developing defenses at what was considered an outlying post protecting Rabaul and an advance staging base for further thrusts into the South Pacific to cut the southern lifeline. A delay in the landing would have provided a better-equipped landing force and more air support. The island was too large to defend and the Americans could have landed at will where they desired. With a better-equipped landing force, they could have cut troop movement and supply routes to reach the airfield area, in much the same manner as they cut supply trails to support operations northwest of the airfield. If nothing else, the Marines, later reinforced by the Army, could have established a lodgment at Aloa Bay, thirty-one miles east of Lunga Point (Henderson Field), where a U.S. airfield was established in November 1942. From there, a landing could have been made at Koli Point nine miles east of Lunga Point. A lodgment at Koli Point, supported by aircraft from Aloa Bay, would have been successful in the same manner as the later lodgments at Bougainville and New Britain. Additionally, the Japanese main headquarters on Tulagi and the other small islets off Florida Island would have been easily neutralized.

Future Japanese plans, curtailed by the June 1942 Midway defeat, saw the 17th Army tasked with seizing New Caledonia, Fiji, and the Samoas in July and August 1942. The operation would have been supported by elements of the Combined Fleet, Second Fleet, and 11th Air Fleet. This plan had been approved in April. The 17th Army forces were completely inadequate for the task, and Japanese estimates of the defenders' strength fell far below what were actually deployed:

ISLAND	JAPANESE ESTIMATE	ACTUAL STRENGTH
Samoa	750	30,000
Fiji	7,500	10,700
New Caledonia	3,000	21,000

The Japanese had only brigade-size forces assigned to seize each of these objectives. With the extended supply lines and increased distances from supporting air bases, these operations could not have succeeded. The operations were cancelled in July 1942.

Sumner Whitten

The failure to capture Guadalcanal would have given the Japanese the opportunity to reinforce and resupply their troops to a much higher degree, perhaps by 100 percent or more. This would have meant that the United States would have lost the only Marine division in the Pacific and that our naval forces would have been driven from the area with heavy losses. It would have permitted the Japanese to more easily build up their forces in New Guinea. That would have resulted in larger and larger Japanese air raids on Australia. The Japanese navy, though seriously depleted, would have by then thoroughly controlled the South Pacific and might have been in a position to blockade Australia and New Zealand. This would have put the British forces in Egypt in jeopardy because most food supplies and manpower for the North African front came from these sources.

If the United States had delayed its Guadalcanal offensive until such time that it had increased its carrier strength, the delay could have lasted up to a year or more, depending upon how long the U.S. Navy would have wanted to wait for additional carriers, battleships, cruisers, and support vessels. Such a delay would have resulted in the Japanese completion of their airstrip on Guadalcanal and would have opened up opportunities for their bombers and fighters. It would have perhaps become almost impossible to drive the Japanese out of the Pacific. Remember that the Japanese fleet was still reeling from the defeat at Midway, much like the U.S. fleet had been reeling after Pearl Harbor. I believe that the American offensive against Guadalcanal found the Japanese overextended for what their supply system could properly sustain.

Keith Allen

After Midway, I don't know that a big Japanese push farther into the South Pacific was likely. Certainly, the Japanese could have bolstered the defenses of Guadalcanal, but they were still at the end of a long supply chain, while the Allies had Australia and New Zealand as large and virtually impregnable bases for South Pacific operations. We also could have bypassed and cut off Guadalcanal, taking other islands in the Solomons or Bismarcks to cut it off.

I doubt the Japanese had the strength to move to Fiji and Samoa at this point. They would probably have mounted a greater effort in New Guinea instead, and might well have taken Port Moresby. I don't think the United States would have done better by bypassing Guadalcanal. Any other obvious target, like New Georgia or Bougainville, would have been contested just as intensely, and would have been much closer to the Japanese air bases at Rabaul.

Jim Hallas

The invasion of Guadalcanal was undertaken on a shoestring, launched over the objections of General Douglas MacArthur and Vice Admiral Robert L. Ghormley, who commanded naval forces in the South Pacific area. MacArthur and Ghormley recommended postponing operations in the Solomons area until U.S. strength—particularly airpower—could be built up. Their reluctance was swept aside by Admiral Ernest J. King, commander in chief of the U.S. Fleet, who insisted it was time to take the initiative against the Japanese. King prevailed and Operation Watchtower went forward.

King's audacity brought some immediate rewards. Though heavy fighting took place on Tulagi and Gavutu, the landing on Guadalcanal itself was unopposed. The 1st Marine Division quickly captured the uncompleted Japanese airfield—soon to be named Henderson Field—as a handful of enemy labor troops fled into the jungle. The Americans would suffer from their slender resources in the weeks and months to come, but seizure of the airfield had come easily, and despite the trials ahead (and they were severe), they would never relinquish it.

While the caution voiced by MacArthur and Ghormley was not unjustified, the advantage of waiting to build up U.S. forces would presumably have been nullified by allowing the enemy the same advantage. In this case, time might have been a greater ally for the Japanese. The Marine landing on Guadalcanal would not have been so easy had the Japanese enjoyed an operational airfield and a garrison of regulars waiting on the beaches. As it turned out, the eventual U.S. success on Guadalcanal came hard—but a success it was. Ironically, the operation also benefited MacArthur in that the Japanese diverted resources from the campaign in New Guinea in their determination to regain

Guadalcanal. Exhausted Australian and American troops on New Guinea were granted a much-needed reprieve and chance to regroup.

Guadalcanal took on tremendous significance as a show of U.S. resolve, a demonstration that the Japanese were far from invincible, and for its attrition of resources (an attrition the United States could better afford). However, an American defeat there would not have meant losing the war. U.S. planners chose to fight at Guadalcanal in an effort to contain Japanese advances and to protect lines of communication to Australia and New Zealand. Had the 1st Marine Division been destroyed or forced to relinquish Guadalcanal, the United States would still have enjoyed an operational presence at New Caledonia, Efate, the New Hebrides, and Espiritu Santo. A Japanese victory at Guadalcanal would have posed problems, but not insurmountable ones. The overall battle would have continued.

Frank Jack Fletcher

The fight for Guadalcanal hinged on several factors, including air superiority, holding on to vital pieces of real estate, and controlling the waters surrounding the island.

Vice Admiral Frank Jack Fletcher, in charge of the three U.S. carriers in the vicinity, informed invasion commanders that the carriers could not support the Marines for more than forty-eight hours. Concerned about refueling and land-based enemy aircraft, Fletcher ordered *Saratoga*, *Wasp*, and *Enterprise* out of the area on August 8–9. When transports and cargo ships also departed due to a lack of air cover, the Marines on Guadalcanal and Tulagi were on their own.[4]

As Fletcher left with his carriers, the IJN struck with heavy cruisers, commanded by Vice Admiral Gunichi Mikawa. In the ensuing night surface battle known as the Battle of Savo Island, the Allies suffered frightful losses, including four heavy cruisers.

Mikawa now had the chance to move against U.S. transports if he wished, but fearing an attack by carrier-borne aircraft as soon as the sun rose, Mikawa retired to Rabaul.[5]

Q: *What if a different commander had led the U.S. carriers early in the campaign for Guadalcanal? If Fletcher had not left with his carriers, what impact, if any, would that have had on the Battle of Savo Island?*

Rear Admiral Frank Jack Fletcher, whose decision to leave with his carriers stirred controversy early in the Guadalcanal campaign.

William Sager

Fletcher pulled out his carriers within forty-eight hours of the invasion of Guadalcanal, although he had indicated to both Admiral Kelly Turner, who was in command of amphibious operations, and General Vandegrift that he would stay for at least seventy-two hours. He communicated to Ghormley that he was running low on fuel and that he needed to retire to refuel. He asked Ghormley's permission to retire and, as I understand it, retired even without Ghormley's permission. Fletcher's decision to withdraw left the Marines with a "bare behind" on Guadalcanal, without air support and without the transports and cargo ships being fully unloaded.

Historians have severely criticized Fletcher for withdrawing his ships. If Fletcher had not pulled out his carriers, there is no question that he could have attacked the Japanese naval force that was withdrawing after the Battle of Savo Island, early the following morning. Fletcher could have found the Japanese going back up the Slot (New Georgia Sound) returning to their base. Those

Japanese ships had no air cover and his carrier planes could have inflicted severe damage. The Japanese feared for the safety of their ships and that's what prompted them to leave the scene after their victory at Savo Island instead of pursuing the American transports, which were essentially unprotected except for a thin line of destroyers. The Japanese could have used their cruisers to sink the American transports and then we would have been in a bad situation. It would have made a great deal of difference if Fletcher, with his carrier task force, had still been on station.

Lex McAulay

Often the West seems to produce the right person at the right place and time, and Fletcher did well there. Whether anyone else would have done better is almost impossible to argue, given that everyone was learning fast. Fletcher had to conserve the carrier force and moving away to do so was correct; to have left it in reach of the enemy at that time would have been wrong. The Battle of Savo Island showed the deficiencies in the U.S.-Allied command and communications procedures, and it was as well that the carriers were not within reach of the Japanese.

Halsey to the Rescue

On October 19, hoping to jump-start operations on Guadalcanal, Vice Admiral William Halsey took charge in the Solomons, replacing Vice Admiral Robert Ghormley.[6] In short order, Halsey sent additional ships, aircraft, and troops to Guadalcanal and inspired the troops by personally flying to the island for a close-up inspection.[7]

Q: *What if Ghormley had not been replaced by Halsey?*

William Sager

Speaking as a grunt second lieutenant, in command of a rifle platoon, it is my view that if Ghormley had not been replaced by Halsey, the Guadalcanal campaign would have continued. It would have taken longer for the U.S. Navy to have secured the island. The naval losses the United States suffered, under Halsey's leadership, would have been considerably greater had Ghormley's leadership continued.

I suppose we have to start with the presumption, as historians have maintained, that Ghormley was never in favor of the invasion of the Solomon Islands, specifically Guadalcanal and Tulagi. While he may not have been in favor of the invasion, I'm not sure he objected to it. His resistance was more passive than anything else. When the American Navy suffered a terrible surface defeat on August 9 at Savo Island, losing four Allied cruisers, Ghormley realized that the U.S. Navy was facing a formidable foe in the Solomons area.

Historically, Ghormley was a good, solid Black Shoe naval officer with an impeccable service record. He was also somewhat timid with his use of the limited resources he had at his command at his headquarters at New Caledonia. Ghormley's timidity, and not his reluctance, was responsible for his recall.

Had he not been recalled, we would have had a longer struggle in the Guadalcanal area. There's no question, in my mind at least, that the Navy, Marines, and the Army personnel on Guadalcanal would have held out until the American production of warships caught up with the losses that Ghormley's naval engagements had incurred in the battles of Savo Island, the Eastern Solomons, and Cape Esperance. In those engagements, the U.S. Navy lost considerably more tonnage than the Japanese navy. Ghormley had great reluctance to risk his limited assets.

As grunts, our greatest criticism of Ghormley was that he never visited Guadalcanal during the entire time the Marines and the Army were located on the island and while Ghormley was at Nouméa, New Caledonia, protected by his remaining naval forces. This contrasts sharply with Admiral Halsey. After Halsey relieved Ghormley upon Admiral Nimitz's orders, Halsey visited Guadalcanal and visited the frontline troops, making himself visible to the infantry and the artillery crews who were bearing the brunt of the battle. Ghormley never visited the Canal and, as far as I was aware, never sent ranking staff to the island. In effect, Ghormley expressed doubt as to whether he could continue to supply the Marines and the Army personnel on the island.

Despite Ghormley's timidity and his failure to properly use his limited assets in an aggressive manner, it is my view that even if Ghormley had not been replaced, the U.S. Navy still would have prevailed over the Japanese. It would have taken longer, and the Marines would have had to tighten their belts a bit to continue to protect Henderson Field. Ground-troop casualties would have risen and morale would have suffered, but eventually the United States would have succeeded, due to America's ability to produce ships and planes.

Plans had already been made to deal with a worst-case scenario. If it appeared that the Japanese were ready to retake Henderson Field, our planes would have been flown off the island to New Caledonia or the New Hebrides, and the Marines would have scattered into the mountains of Guadalcanal to fight a guerilla battle. Guerilla warfare would not have been terribly successful on Guadalcanal. Guerilla warfare was better suited for a larger land mass, like the Philippines or mainland China, where you can move your guerillas around and get the support of the local population.

Halsey's aggressiveness and his willingness to risk warships, however, accelerated America's victory on Guadalcanal and reduced the total number of casualties that otherwise would have been suffered by the U.S. forces on the island.

Douglas V. Smith

I am convinced that this was one of the most important decisions Fleet Admiral Nimitz made during the entire war. I say this as someone who is not a big fan of Admiral William F. "Bull" Halsey. Vice Admiral Ghormley, as commander of the South Pacific Area (COMSOPAC), was faced with an almost impossible situation from the time he assumed his command on June 19, 1942, through the early stages of the Lower Solomons campaign. He was overwhelmed by a tenuous and deteriorating situation, highlighted by the biggest disaster in the history of the U.S. Navy in the Battle of Savo Island, and was defeatist and fatalistic regarding the possibility of success against the Japanese onslaught to reestablish control on Guadalcanal. On October 15 he went to the extreme of declaring his forces "totally inadequate" to repel a major Japanese offensive aimed at wresting control of Guadalcanal, which his intelligence predicted would come around October 23, in a message to Admiral Nimitz in Hawaii. When any commander is resigned to defeat, his command responds accordingly. Nimitz had the extremely hard choice of acting immediately and taking the chance of losing all continuity by changing his operational commander so close to a major action, or facing unacceptable defeat with a commander who was clearly not up to the challenge. If Vice Admiral Ghormley had remained in his position at that critical juncture in the Pacific War, I am absolutely convinced that the American operations to retain their foothold on Guadalcanal would have failed, regardless of the timely arrival of the 164th Army Infantry Regiment to reinforce the Marine situation on Guadalcanal on October 13. "Bull" Halsey's leadership style was, "Damn the consequences—full speed ahead." While that on occasion got him into trouble, it was exactly what was needed to rally those in his charge to a seemingly impossible task. This, in my view, was Admiral Halsey's "finest hour." By sheer strength of conviction and positive attitude, he turned the tide and achieved victory when lesser men would have failed.

Japanese Mistakes

The Japanese made every effort to eradicate the Americans on Guadalcanal.

On August 21, about one thousand men led by Colonel Kiyono Ichiki delivered the first blow, a suicidal assault against Marines dug in along the Ilu River. Very few attackers survived; Ichiki himself later committed hara-kiri. In September, a Japanese offensive aimed at capturing the airfield near Lunga

Beached Japanese transports at Guadalcanal, as an SBD bomber (foreground) flew by on November 16, 1942. Could the Japanese have found a way to push the Americans off Guadalcanal?

Point failed; in October, the Americans stopped another thrust, this time by an entire division.[8]

At sea, the Imperial Japanese Navy triumphed in several surface battles, most notably Savo Island on August 9, when Vice Admiral Gunichi Mikawa led five heavy cruisers, two light cruisers, and a destroyer against American and Australian warships. Savo cost the Allies more than one thousand sailors, but instead of advancing on enemy transports, Mikawa retired.[9]

In October and November, Japan used capital ships to blast Henderson Field; despite some scary moments, the Americans held their positions. They also inflicted heavy casualties on Japanese reinforcements on November 14, when planes from *Enterprise* and Henderson Field sank six transports, damaged another, and repeatedly bombed four others, eventually forcing them to beach. Japan also lost the battleships *Hiei* and *Kirishima* after surface battles on November 13 and November 15. Through it all, the Americans remained firmly entrenched on Guadalcanal.[10]

Q: *Could the Japanese have used different tactics at Guadalcanal?*

Jon Parshall

The only thing that was going to save the Japanese position, temporarily, on Guadalcanal was a prompt recognition that this was a major American effort, followed by absolute priority being given to Guadalcanal, and the immediate commitment of divisional-sized ground forces supported by every heavy unit in the IJN. Historically, the Japanese placed themselves in a position of constantly trying to raise the ante a little bit at a time, not wanting to be sucked into too large a fight. The result was that they were always a bit behind the game. Only by pushing all the chips into the pot up front could they have possibly avoided this. Unfortunately for them, Guadalcanal's very nature made it very difficult to discern that this was, in fact, the decisive battleground of the Pacific War. It had none of the characteristics of such a battlefield, at least according to the IJN's doctrine. As such, they reacted more slowly to the developing menace than they ought to have.

Keith Allen

Certainly, the Japanese should not have conducted the campaign in such a piecemeal fashion. I'm not sure I blame them for not committing ships like

Yamato and *Musashi* in the confined waters of the Solomons in a night engagement (although it worked for us on November 14–15), but they definitely could and should have made a more coordinated air-ground-sea push, committing their carrier forces and a much stronger ground component, and making an all-out effort to secure air superiority in the Solomons. They had ground troops available, and in total committed the equivalent of about three divisions during the campaign. They could have pushed a large transport force down the Slot with heavy surface cover, and committed the carrier force, along with strong submarine patrols between the Solomons and the New Caledonia/New Hebrides area. They also could have reinforced their land-based air on Rabaul.

In addition, they should have used their submarines to attack Allied lines of communication to Australia, New Zealand, New Caledonia, the New Hebrides, and the Solomons.

Jim Hallas

The edge between victory and defeat at Guadalcanal was far narrower for both sides than is popularly recognized. Since it happened, an American victory seems inevitable in retrospect, but that was far from the case. At least twice during the course of the campaign General Alexander A. Vandegrift, commanding the 1st Marine Division, alerted his staff officers to make plans to continue the fight from the island's interior should the Marines be overrun.

This was not theater. In the initial weeks, the Japanese controlled both air and sea. During the Battle of Savo Island on August 9, the U.S. covering force lost four heavy cruisers and a destroyer along with nearly 1,300 men killed. The Japanese were hardly scathed. Luckily, Vice Admiral Gunichi Mikawa apparently did not recognize the scale of the landings and failed to attack the American transports. Still, the disaster persuaded Admiral Fletcher to withdraw his ships, leaving the 1st Marine Division hanging on by its fingernails.

Some U.S. naval officers believed Vandegrift was doomed. Fortunately for Vandegrift, the Japanese made a number of fatal mistakes, beginning with underestimating the size of the U.S. force on Guadalcanal. Colonel Kiyono Ichiki kicked off the parade of failures by rashly leading about one thousand men directly into American machine guns on the night of August 20–21. The force was virtually annihilated.

The Japanese subsequently landed well over a division on Guadalcanal, but were unable to overwhelm the Marines. Japanese efforts were hampered by the horrendous terrain, which contributed to unit disorganization, exhausted the troops, who had to slog through miles of jungle, and hindered efforts to bring heavy supporting weapons into play. Communications failures contributed to disorganization. Poorly coordinated, piecemeal attacks by lightly equipped infantry were cut to pieces by dug-in Marines, even when the Japanese were able to bring numerical superiority to bear against a chosen point in Vandegrift's perimeter.

Even so, the Japanese came very close to breaking through and capturing Henderson Field. The Marines on Edson's Ridge, blocking access to the airfield, were nearly overrun on September 12–14. One group of Japanese even made it to Vandegrift's command post. Major assaults were also launched on October 23 and 24, and at one point a small number of Japanese managed to penetrate the U.S. perimeter before being driven back.

Considering how close the Japanese came to success despite all of the obstacles—natural, those presented by the U.S. Marine Corps, and those that were self-inflicted—it is likely that a more judicious approach, with better intelligence and coordination, could very well have forced the Marines into the jungle. A speedier, coordinated mass landing by more troops—say, two full divisions—backed by a powerful naval force might have ended the American effort almost before it began.

Q: *What if the Japanese had used more men in early assaults on Henderson Field?*

Jon Parshall

Where would those men have come from, and how would they have gotten to Guadalcanal? It's not like the Japanese had several divisions' worth of ground troops just sitting around twiddling their thumbs while the Ichiki detachment went down to Guadalcanal and got slaughtered. This points out a central problem with their whole campaign in the South Pacific—they had run it on a shoestring at the end of a supply line that was about four thousand miles long. There weren't any substantial forces in the area, and what forces were there were committed to the New Guinea campaign. In fact, it's not really possible to discuss Guadalcanal without also contemplating the likely impact on New Guinea, because the Japanese really were playing a zero-sum game in the southwest Pacific in terms of troops. They were trying to capture too many objectives with too few troops. That's why Ichiki was sent down to Guadalcanal with what he had—there weren't any more troops to be sent, and the transports weren't available to send them in any case.

William Sager

As some military historians have said in the past, it wasn't so much the Marines that defeated the Japanese as it was the jungle. During the Battle of Bloody Ridge, on September 12–13, 1942, the Japanese force of approximately 4,500 came in from the south, attacked the Marine Raider battalion holding the ridge south of Henderson Field, and almost dislodged the Marine battalion from the ridge. What if the Japanese commander had not attacked in the ridges south of the airfield? What if instead the Japanese had attacked across the grass field from the Ilu River west of the airfield, rather than merely launching a diversionary attack in that area against the 3d Battalion, of the 1st Marines, September 12–13? Some military historians have raised this possibility in the past, including William H. Bartsch, who stated that if General Kiyotake Kawaguichi had focused his attack on the 3d Battalion line, in a major assault from the east, rather than attacking the more heavily defended Bloody Ridge from the south, the history of the Guadalcanal campaign might have ended then and there. I'm not as pessimistic. Admittedly, General Kawaguichi's substantial force would have overrun the line of Marines of the 3d Battalion and would have reached the airfield, about a quarter of a mile behind the 3d Battalion's lines. The Japanese would have undoubtedly destroyed the planes at the airfield. It's unlikely that General Kawaguichi would have been able to hold that position without substantial support from the Japanese navy, and without Japanese planes arriving to upgrade his position. General Vandegrift would have been hard-pressed to assemble a force of Marines sufficient to retake the airfield. Although the numbers might have favored Vandegrift, the location of his troops among the perimeter defense would have caused problems. Even so, at this particular juncture, Vandegrift still had five battalions of Marines, supporting troops, supporting tank battalions, and also heavy artillery, which Kawaguichi did not have. The Marines could have carried the day and chased Kawaguichi's forces off that airfield. It would have been a tough fight. We had no reserves, all of our ground troops would have been committed, and we would have been working without air cover, since presumably Kawaguichi would have been in command of the airfield. To have taken advantage of the airfield, Kawaguichi needed support, ammo, fuel, and ground crews. I'm not sure the Japanese would have been able to find a way to bring in those resources to Kawaguichi.

Q: *What if the Japanese had used Yamato and Musashi in night surface battles around Guadalcanal?*

Jon Parshall

This, I think is one of the biggest variables in the battle, because if the Japanese had made a concerted effort to shut down Henderson Field with surface gunfire, I think they could have made a definite impact on the battle. Any Guadalcanal vet that I've ever talked to who survived "The Bombardment" in October will tell you that it was a profoundly depressing event, and a very, very scary time for the Americans. Had the Japanese been able to do that a few more times, and bought enough time to bring their transports in, the ground battle could have definitely gone the wrong way for the Marines.

The reason they didn't, I'm convinced, is two-fold. First of all, Japanese naval doctrine placed a premium on preserving the battleships of Battleship Divisions 1 and 2 (which included *Yamato* and *Musashi*) for usage in a decisive sea battle. The fast battleships, such as *Hiei* and *Kirishima*, were part of Battleship Division 3, and while they were nominally a part of the Japanese battle line, it was understood that their protection was insufficient to let them really work with the battle line. This is why they had been freed up to act as carrier escorts. And this is also why they were considered "expendable" enough to risk in the restricted waters off of Guadalcanal in support of operations that didn't exactly fit the mold of a decisive battle.

The second factor was almost certainly fuel. Contrary to the common image of Truk being this sort of mysterious Japanese mega-base, in fact it was really a backwater. Its fueling facilities were minimal, and fuel was often dispensed from oilers anchored in the lagoon. The same was true of its repair and other logistical facilities. Committing Combined Fleet's battleships to a concerted effort to shut down Henderson would have meant very many sorties back and forth from Truk down to Ironbottom Sound, each of which would have consumed thousands of tons of fuels. At the same time, however, fuel stocks were dwindling at the naval bases in the Home Islands—this is mentioned explicitly in some of the Japanese sources. Thus, it might have been possible to do this, but it would have been seen as both highly risky to the very valuable heavy capital ship units within the fleet, as well as very expensive in terms of logistics.

H. P. Willmott

I do not believe that *Yamato* and *Musashi* would have made any real difference, certainly not the latter: she was not completed until August 1942, and with normal lead times this would mean that she could not have been off Guadalcanal. The situation with *Yamato* was slightly different, but I have always been of the view that had the American and Japanese battleships fought then the result would have been an overwhelming victory for the Americans. Of course, the weakness of that argument is that one assumes 1944 levels, specifically radar, in 1942 and that was not necessarily true, but I think in sheer volume of firepower the American battleships, and most obviously the *Iowa* class but I think also the *North Carolina* and *Washington* classes as well, would have overwhelmed any Japanese battleship(s) that appeared off Guadalcanal.

I also genuinely believe that Yamamoto has been much overrated. That claim about waking the United States: nice, but no one can state when and to whom he made the remark, and my guess is that this is postwar make-believe when the blame was being passed to the Rikugun (the Imperial Japanese Army Academy, where officers were trained) and Yamamoto was dead. I think the Japanese naval conduct of the Guadalcanal campaign was abysmal, and the conduct of the carrier actions in the Eastern Solomons in August and in Santa Cruz in October was worse.

The Japanese problem is very simply defined: they were always going backward. Once the Americans were ashore then the Japanese were always trying to play catch-up, and once the American aircraft were on Henderson Field then the Japanese defeat was there in the making. What the Japanese had to do was concentrate all their aircraft and a corps and make the one massed effort, though, as at Midway, their carriers would have been caught with attention divided between the neutralization of Henderson Field and the search for American carriers. But my real point would be that if the Japanese had prevailed on Guadalcanal, what next? The Americans would have bypassed them and gone on to Bougainville.

What never seems to get the attention it merits is the inconsistency of Japanese intent: the holding of fixed positions in a war of movement.

Santa Cruz

Guadalcanal featured two big carrier battles. The first, the Battle of the Eastern Solomons on August 24, resulted in the United States temporarily losing the services of *Enterprise*, which sustained three bomb hits, while the IJN lost the light carrier *Ryujo*.[11] In ensuing weeks, Japanese subs damaged *Saratoga*, knocking her out of action for three months, and sank the *Wasp*.[12]

Then came the second big carrier battle on October 26, the Battle of the Santa Cruz Islands. While the Americans damaged carriers *Shokaku* and *Zuiho*, the Japanese slammed bombs, torpedoes, and flaming planes onto the *Hornet*. Eventually, American and Japanese torpedoes sent the U.S. carrier to the bottom.

Enterprise, heavily damaged at Santa Cruz, slipped away for repairs at New Caledonia and was soon ready to re-enter the fight for Guadalcanal.[13]

Q: *What if the Japanese had sunk* Enterprise, *in addition to* Hornet, *at Santa Cruz?*

John Lundstrom

Halsey was reckless in the Battle of Santa Cruz in October 1942. He fought at the edge of his land-based air search. He was near the top of his umbrella when it came to scouting for enemy ships. If the Japanese had been able to sink the *Enterprise* in the battle, and all of those American planes had been lost, you really have to wonder what the result might have been at Guadalcanal. I'm not sure the Americans could have held the island without any support from U.S. carrier planes. In fact, the Japanese believed that they had knocked out all of the American carriers around the island and that's why they sent the reinforcement convoy down to the island in November. The Japanese had thought that they had won. If those reinforcements had gotten through, they would have bolstered the Japanese forces on the island.

Without the *Enterprise*, the Americans would have had to wait for the *Saratoga* to return and at that stage, what could one U.S. carrier have done? If the *Enterprise* had gone down along with the *Hornet* at Santa Cruz, it would have been a tremendous blow to American morale.

Alvin Kernan

The Japanese did not win the battle, but it is difficult to know where they made a mistake. They had scouted effectively, except for the delay in the first report, launched their attacks as quickly as possible, found their targets, and attacked, as always, skillfully and courageously. Their attacks were much more effective than the attacks from the American carriers, which as usual broke up on the way to the target, got lost, and hit two enemy carriers almost by chance with very small portions of their total strike group. The Japanese would have sunk both American carriers had they not dropped the ball in strikes two and three. Strike one on the *Hornet* went off brilliantly, but when Nagumo's dive-bombers of his second strike group reached the *Enterprise* at 10:15 a.m., his Kate torpedo planes did not show up until twenty minutes later—some navigational delays presumably. The Japanese successes in the war from Pearl Harbor on had come from combined attacks in which dive-bombers occupied the targets while the torpedo planes came in and delivered the heavy ship-sinking blows. They had just used these tactics brilliantly on the *Hornet*, which lay burning on the horizon, sinking not from the bomb hits but from two torpedoes.

It needs to be remembered that Japanese torpedoes were superb weapons, in submarine, shipboard, and aerial delivery. The tactics for delivering them aerially were equally superior at Pearl Harbor, Midway, and Santa Cruz. The American aerial torpedo, the Mark XIII, was a dud, and the delivery methods were suicidal and ineffective. When Japanese carriers were disabled as at Midway and Santa Cruz, they were put down with dive-bombers, our primary working weapon. But when American carriers went down it was from torpedoes, and the failure of the Japanese torpedo planes to attack simultaneously with their dive-bombers in the attacks on the *Enterprise* doomed them to failure. Both torpedo attacks on the *Enterprise* went in without the dive-bombers. The ship could give all its attention to avoiding the torpedo tracks, which it did very successfully.

Had the Japanese coordinated their dive-bombers and torpedo planes they would almost surely have put the *Enterprise* on the bottom. Since the war was eventually won by America overwhelming Japan with its war production and trained manpower, it cannot be claimed that the absence of all American carriers from the Guadalcanal area at the end of 1942 would have cost the war, but it would certainly have lengthened it substantially. The two Japanese carriers still operational, the smaller *Junyo* and the larger unharmed *Zuikaku*, would have been free to operate off Guadalcanal, with whatever air groups they could cobble together at Rabaul and Truk. They could have prevented our surface ships from entering those waters to contest Japanese surface ships coming down the Slot, to reinforce the army on the Canal and to bombard Henderson and at last put it out of commission. Under constant attack and without air support, the Marines would have been starved out eventually and Japan would have been free to continue its expansion to the south and the east, cutting the supply line to Australia and the Allied army forces there.

There was a lot riding on American air control of the area around Guadalcanal and, at the end, the *Enterprise* was the key to it. When she came back on station in early November, no matter how battered, and when the repaired *Saratoga* joined her shortly afterwards, American control of the air around Guadalcanal was assured.

The Japanese carrier force did not reassemble to challenge the American ships at Guadalcanal, and this was the end of the carrier-to-carrier battles until the summer of 1944.

Summary

Whereas the Japanese lost the Battle of Midway in minutes, it took them nearly six months to acknowledge defeat on Guadalcanal.

Japan could point to missed opportunities, like Mikawa's failure to press his advantage after drubbing Allied cruisers off Savo Island in August. There were also the ground assaults in September and October that nearly broke U.S. lines. As some of the experts noted, if the Japanese had been quicker to recognize that Guadalcanal was a decisive battleground, they might have coordinated their resources more effectively.

For the Americans, Guadalcanal was a campaign of attrition. They lost three carriers—*Hornet*, *Saratoga*, and *Wasp*—and at one point were down to just the *Enterprise* remaining. In the end, the Americans held onto Henderson Field, harried enemy reinforcements, and forced Japan to evacuate Guadalcanal, pinning another costly defeat on the Japanese Empire.

Nine

BEGINNING OF THE END: 1943–1944

IN JUST EIGHT MONTHS, Japan's prospects had taken a remarkable turn: Ambushed at Midway and exhausted on New Guinea and Guadalcanal, the Japanese were now firmly on the defensive.

If Japan could not find a way to slow down the Americans, its eventual defeat seemed likely, just as long as the Americans were willing to suffer the heavy casualties that were sure to come from taking the war to the Japanese.

Ambushing Yamamoto

Admiral Isoroku Yamamoto, commander in chief of the Combined Imperial Fleet, entered the war with full knowledge of America's industrial might. Educated at Harvard, Yamamoto also served as a naval attaché in Washington and had misgivings about fighting America.[1]

In April 1943, U.S. code-breakers struck again. This time, they learned that Yamamoto planned to visit Bougainville, a Japanese base within reach of Army P-38 fighters on Guadalcanal. Admiral Nimitz approved an ambush.

On April 18, P-38s shot down two Betty bombers, including Yamamoto's plane, killing the man who had been so instrumental in wrecking U.S. battleships at Pearl Harbor some sixteen months earlier. Yamamoto's replacement, Admiral Mineichi Koga, who had served in the past as vice chief of the Navy General Staff, would now have the unenviable task of trying to stop the Americans.[2]

Q: *What if Yamamoto had continued to lead the Combined Fleet beyond April 1943?*

Admiral Isoroku Yamamoto. How much did his death in the spring of 1943 hurt the Japanese war effort?

Anthony P. Tully

This seems to be a "what if" whose importance is easy to overestimate, rather than underestimate. From Yamamoto's mixture of gifted and flawed strategic responses since Pearl Harbor, and especially during the Guadalcanal campaign, it does not obviously follow that his successor's decisions and strategies were markedly inferior.

In my opinion, Admiral Mineichi Koga's role as commander in chief of Combined Fleet has been underestimated and obscured both by the ultimate total failure of his efforts, and his untimely death.[3] Nonetheless, given the resources at his disposal and the limitations placed upon him, the plans devised for Combined Fleet, and especially the A-Go operational outline, seem an effective and potentially successful use of his varied but limited resources of land, air, and surface forces.

The question of Koga's competence has retroactive implication regarding the survival of Yamamoto. If it can be shown that Koga's use of the fleet approximated the likely use and strategy Yamamoto would have done, then the latter's death in strategic terms becomes less decisive.

When Yamamoto died, the I-Operation—the big aerial offensive in the Solomons—was entering the final stages of another disaster. It is not obvious or apparent that his responses or alternatives to the invasions of the Marshalls in the fall of 1943, or that of Bougainville, would have or could have been much different. On the other hand, one should not in the least underestimate the huge psychological factor and blow to morale on the Japanese side, and corresponding boost to morale on the Allied side, of the ambush. Isoroku Yamamoto was seen by many on both sides as the central genius and architect behind Pearl Harbor, and his elimination was huge. But whether the Japanese navy would have followed a radically different strategy in fall 1943 and leading up to the Marianas seems somewhat doubtful.

The big question mark is whether perhaps lingering guilt or a sense of atonement for the Midway failure would have led Yamamoto to lead the next decisive battle in person, perhaps even sooner than at the defense of Saipan, perhaps during the Solomons campaign at a time of his choosing. But the logistical realities dictating Japanese action in the fall of 1943 would not have been easy to overcome, and in all likelihood Yamamoto would have come up with strategies not dissimilar from Koga's.

Jay Stout

There is a growing tendency among World War II enthusiasts and amateur historians to ascribe more greatness to Fleet Admiral Isoroku Yamamoto than he merits. Some even posit that the war might have ended differently had he not been shot down and killed by a skillfully led—but undeniably lucky—group of Guadalcanal-based Army P-38s.

The fact is that Japan had virtually no chance of winning the war by the time Yamamoto was killed. The pivotal battles at Coral Sea and Midway nearly a year earlier had ended in favor of the United States. And the Japanese had just lost the great struggle at Guadalcanal; this was the first campaign that pitted the combined combat arms of both nations against each other over a protracted period. From its hard-won bases at Guadalcanal, the United States was expanding its influence across the region. Indeed, Yamamoto was killed while touring bases that he was desperate to defend against the burgeoning American advances.

It is important to note that the successes the United States had achieved to this point in the war had been won well before the nation had come even close to "hitting its stride." It didn't matter how clever, resourceful, or hard-working Yamamoto was, the reality was that Japan never could have matched America's ability to produce men and matériel.

A simple reference metric—one of many—was aircraft production. During 1942, the year the United States prevailed at Coral Sea and Midway, the nation produced only 47,675 aircraft. In 1943, it produced 85,433 aircraft. From the time of the Pearl Harbor attack until the end of the war the United States manufactured more than 275,000 aircraft of increasing complexity and capability. In contrast, Japan produced only 66,000 aircraft of largely inferior quality and utility. Japan's relative shortcomings in producing other war material, and in recruiting and training competent soldiers, sailors, and airmen, were even worse.

So, it wouldn't have mattered if Yamamoto had lived—and had additionally been the greatest strategist and tactician that had ever lived—Japan simply couldn't have produced the resources he would have needed to win a prolonged conflict with the United States. And he had said as much even before Japan started the war. In fact, from the time of his death in April 1943, there was no one in history who could have led Japan to victory against the United States.

And, in truth—the industrial capacities of the United States and Japan aside—Yamamoto was a talented strategist but he was hardly infallible. His sneak attack on Pearl Harbor was a well-planned but incompletely executed success. The battle he planned and lost at Midway had been too complex. Admittedly, his forces had suffered miserably bad luck during that fight, but genuine mistakes were made, and the Imperial Japanese Navy slunk away as a gutted mess that would never recover. Finally, Yamamoto failed to prevail against critically stretched American forces at Guadalcanal.

American leaders proved to be his match and more.

So then, although Yamamoto's death was a blow to Japan's morale, and although his loss created confusion and left a void in Japanese military circles, it meant little in the context of the outcome of the war in the Pacific.

Jim Hallas

As the main impetus behind the successful surprise attack on Pearl Harbor, Admiral Isoroku Yamamoto gained a reputation as an imaginative and brave commander. A national hero in his own country, his name is also familiar to many Americans more than sixty-five years after his death (thanks in part to the popular movie *Tora! Tora! Tora!*).

His exalted reputation among the Japanese is understandable. Ironically, considering the number of American servicemen who fell victim to his stratagems, Yamamoto also seems to be viewed today with considerable sympathy and admiration in the United States. This may be because historical accounts of the admiral tend to emphasize his familiarity with this country, his admiration for many things American, and his known reluctance to go to war with the United States. Anyone who has delved even superficially into the naval war in the Pacific recognizes Yamamoto's famous prediction that in the event of war with the United States, "We can run wild for six months or a year, but after that I have utterly no confidence. I hope you will try to avoid war with America."

Of course, it is easier to be sympathetic in today's world where Japan is an ally rather than an enemy. Americans of 1943, quite understandably, were less enamored of the admiral. When the opportunity arose to ambush/assassinate him in the air over Bougainville in April 1943, the mission was seen as a chance to remove a dangerous and resourceful enemy and—some say—exact a measure of revenge for Pearl Harbor. That mission not only succeeded, it became one of the most legendary stories of World War II.

But what were the practical ramifications of that success?

The most immediate effect of the admiral's death was to damage Japanese morale while raising that of the Allies. The architect of Pearl Harbor was dead. The cunning and charismatic leader of the Combined Fleet no longer posed a threat to Americans heading into harm's way.

The longer-term effects of Admiral Yamamoto's death probably turned out to be less significant than his reputation might have indicated. By April

1943, one Japanese admiral—no matter how talented—was not enough to alter the tide turning against Japan. The Japanese defeat at Midway had not only diminished Yamamoto's influence, it had set the United States on the offensive. Yamamoto's prophecy was in the process of being fulfilled. The Japanese navy had run rampant for the first six months of the war, but now had entered a period of attrition—one the United States with its vast industrial base could well afford.

Given the realities of the situation, there is no reason to believe that Admiral Yamamoto, had he lived, would have fared any better in 1944–45 than did his successors, Admiral Mineichi Koga or Admiral Soemu Toyoda. Yamamoto's vision for a decisive naval victory was already out of reach when he died and only became increasingly impossible afterwards as American power and numbers continued to grow while Japanese capabilities withered away.

Admiral Koga acknowledged the bitter truth shortly before his own death in 1944, when he remarked to his chief of staff, "Yamamoto died at exactly the right time."

Tarawa

To win in the Pacific, the Americans had to be willing to accept heavy casualties. While they could bypass strongholds like Rabaul, other Japanese-held islands with important airfields had to fall.

In November 1943, as the Allies pressured New Guinea and Bougainville, the U.S. Pacific Fleet drove into the central Pacific, targeting Tarawa, with its valuable airfield, and Makin in the Gilbert Island chain. Both were tiny, flat atolls with treacherous coral reefs, offering little cover for the invaders.[4]

Makin fell in four days. However, the struggle for Betio and its airstrip, in the southwestern corner of the Tarawa Atoll, was horrendous. At two miles long and no more than nine hundred yards across, Betio was a tough nut to crack. Nearly 4,800 determined defenders awaited the Americans, many in bombproof bunkers lined with steel and concrete. The assaulting troops also had to contend with coastal defense guns, artillery pieces, machine guns, and a four-foot-high coconut-log seawall.[5]

On November 20, after a pre-invasion bombardment, Marines on board LVTs (landing vehicle, tracked) and LCVPs (landing craft, vehicle and personnel) headed toward shore. Heavy fire knocked out LVT after LVT; thanks to a low tide, the LCVPs (Higgins boats) ran aground on Betio's coral reef. The Japanese had no shortage of targets.[6]

As casualties mounted, Marines slowly began to eliminate shore batteries and pillboxes, slugging it out yard by bloody yard. Betio's commander, Rear Admiral Keiji Shibasaki, perished in the fighting.

Declared secure on November 24, Betio claimed the lives of more than one thousand Americans and all but seventeen of the Japanese defenders. Admiral Spruance would later defend the decision to invade the Gilberts and the tactics used at Tarawa.[7]

Q: *What if there had been better intelligence about tidal conditions at Tarawa or the Americans had done a better job softening up Japanese defenses? What if the Japanese commander on Betio had not been killed in the fighting?*

Jim Hallas

Intelligence for the Tarawa landings was actually very good. ULTRA intercepts and aerial photographs were especially valuable, revealing details on the

Marines capture a Betio Island pillbox by frontal assault. Was the Tarawa invasion necessary?

enemy garrison, its numbers and composition, and the number and types of defense positions. The 2nd Marine Division's enemy situation map distributed in late October was later found to be nearly 90 percent accurate. That's surprisingly good intelligence for an operation that is often dismissed as an unmitigated mess from start to finish.

Intelligence was also aware, from interviews with former residents and other data, that the water over the reef might be low, but expected the tide to rise sufficiently to allow boats to make their way in to the beach. By a tragic coincidence, the date chosen for the Tarawa landings was one of only two days in 1943 when the moon's apogee coincided with a neap tide. This resulted in a tidal range of inches instead of feet. These unusual circumstances were not fully understood for nearly forty-five years, so it seems unfair to lay great blame on U.S. intelligence resources of the day.

It is generally taken for granted that had planners been aware of this unusual event, lives would have been saved by rescheduling the operation for a more favorable tide. The ability to cross the reef in boats following the initial LVT assault would have put men and matériel on the beach more quickly and in greater numbers, overwhelming the enemy, or so the reasoning goes. The slaughter on the reef could have been avoided.

However, Colonel Joseph H. Alexander in his definitive study of the battle, *Utmost Savagery: The Three Days of Tarawa*, questions that conventional wisdom. He points out that higher water would have slowed the LVTs in the initial waves, making them more vulnerable to anti-boat fire. Using their tracks on the coral, the vehicles were three times faster than their speed of three to four knots when afloat. Similarly, had there been enough water for the slow-moving conventional landing craft to get over the reef, the Marines might merely have been killed by the boatload by heavy guns instead of individually as they waded ashore.

Finally, Alexander observes that a higher tide would have erased that small shelf of ground below the seawall—the only refuge for the battered assault waves during those first hours. Certainly many of the wounded would have drowned. Alexander also points out that despite the notoriety given to that long, terrible wade in to the beach, 80 percent of the casualties suffered by the Marines at Tarawa occurred on land, not in the water, evidence that the impact of the low tide on personnel losses has been overstated.

On the other hand, had U.S. intelligence recognized the extent of the rare low tide, planners might have been able to obtain more LVTs for the landing force (perhaps transferring some of those slated for the less-contested Makin operation). As it was, there were only enough LVTs to transport the first three waves ashore, and many of the tracked vehicles were soon knocked out. Nevertheless, the Japanese were stunned by the ease with which those initial waves scuttled across the reef and established a toehold on the beach. More LVTs would have saved Marine lives and helped get more men ashore faster and in a better state of organization than was the case.

Casualties could also have been reduced by a longer and more effectively delivered naval bombardment and better air support. Planners for the 2nd Marine Division proposed a naval bombardment lasting several days, as well as establishment of an artillery firebase on nearby Bairiki Island prior to the landing. They got neither. Fearing interference from Japanese naval forces and insisting on retaining the element of surprise, Admiral Nimitz allowed only three hours of preparatory fire.

It was not nearly enough. The bombardment was visually impressive, but did little damage to the five hundred pillboxes and defensive positions on Betio. Those guns survived to inflict horrendous casualties on the landing force. The death toll was increased by a last-minute failure of coordination between naval gunfire and air strikes. As the landing craft headed for shore, the naval bombardment was lifted in order to allow air strikes to take place. The planes didn't show up and the Japanese had over twenty minutes to emerge from shelter, man their guns, and prepare for the landing.

On a day when everything possible seemed to go wrong, the Marines did have one stroke of good luck: the very capable Japanese commander at Tarawa, Admiral Keiji Shibasaki, and his staff were killed by naval gunfire during the afternoon when the admiral vacated his command bunker so it could be used to house the wounded. Had Shibasaki lived, he presumably would have directed an organized counterattack against the beachhead that night when the Marines were barely hanging on. That counterattack would have had an excellent chance of pushing the Marines back into the sea.

Don Allen

Estimates made of the number of defenders were very accurate, using the number of over-the-water latrines that had been built. The calculated number was

4,840 and the actual count came to 4,836. Aerial reconnaissance revealed nearly all of the larger weapon emplacements, and considering there was no way to gather human intelligence from the atoll, the Navy had a pretty clear picture. The biggest intelligence failure had to be disregard of the neap tide warning from a New Zealand officer who had lived on Betio for several years.

Had they been able to offload on the beach, there still would have been tremendous destruction of landing craft due to the accuracy and effectiveness of the Japanese dual-purpose guns along the landing beaches. By the same token, if the tide was up there would have been little or no seawall for protection once they offloaded. The entire island could be readily swept by gunfire, and they would have been hit as the ramps went down, like at Normandy, but from a much closer range.

If they had accepted the neap tide prediction and offloaded at the 700-foot reef's edge, as some of the LCVPs did, the outcome may not have been a lot different. It was like an arcade game for the defenders, hitting the Marines as they waded the incredible distance to the safety of the seawall. The amtracs were the biggest boon to the landing in this situation, getting the men to the beach faster, but were still very vulnerable.

As for the naval bombardment, the highest point on Betio is, perhaps, ten feet above sea level. The Navy soon discovered that some of its shells were ricocheting off the island and landing among our ships on the other side. The downrange ships were soon moved, but many of the shells still bounced off the island. Plunging fire would have been much more effective, especially against the coconut-log and sand-reinforced bunkers. A full day of bombardment before the invasion would have been demoralizing for the defenders and may have reduced more of the defending guns. The pre-invasion D-day mission by B-24s from Funafuti did not materialize.

Q: *Could Tarawa have been bypassed? Were the lessons learned worth the casualties?*

Jim Hallas

Defeat at Tarawa would have forced the United States to question the legitimacy of its developing amphibious doctrine. Abandonment of that doctrine would have dramatically altered the course of the war in the central Pacific—perhaps even ending the Navy's island-hopping strategy in favor of throwing all resources into MacArthur's advance through the southwest Pacific. Thanks to the decapitation of the Japanese command structure, and the effect of the bombardment on Japanese communications, the counterattack that might have carried the day for the Japanese never took place.

After the war, General Holland M. Smith proclaimed in his very controversial memoirs that Tarawa had been unnecessary. More reasoned thinking indicates otherwise. The Gilberts were a necessary prelude to the strategically valuable Marshall Islands. Betio's airstrip was necessary for photo reconnaissance and air strikes on the Marshalls and to provide land-based air support. Left in Japanese hands, Betio and its airstrip would have presented a threat to U.S. lines of communication and shipping.

Was it worth the casualties? One dead Marine is one too many. But U.S. amphibious doctrine had to be tested somewhere. That somewhere happened to be Tarawa. Mistakes were made; lives were lost that might have been saved, but the basic doctrine was vindicated. Thanks to Tarawa and the lessons learned, the Marshall Islands campaign was a stunning success; thanks to Tarawa, two Marine divisions at Saipan stormed across the reef in more than seven hundred LVTs. Tinian, Guam, Peleliu, Iwo Jima, and Okinawa all were won thanks in part to those early lessons at Tarawa.

Don Allen

Medium and heavy bombers require a land airstrip. The *Hornet*'s attack on Japan was a one-time deal. Our bombers from the nearest airstrip seven hundred miles south could not safely reach the Marshall Islands, the next target in line. Tarawa had a ready-made strip that could handle medium bombers and fighters, and which was lengthened to accommodate heavies. A second, better and bigger strip was built after the battle on the eastern island of Bonriki, and is still in use today.

Technically, just about any of the island battles could have been bypassed, allowing the defenders to starve by blockade. Tarawa proved to be not only useful as an airstrip, but useful in the lessons learned from the operation. Tarawa revealed the need for underwater demolition teams (forerunners of today's SEALs—sea, air, and land maritime Special Forces) to neutralize offshore mines and obstacles. Types and designs of defensive positions were duplicated in training areas back home, and proved useful in subsequent assaults.

Tarawa was truly an experiment in target preparation, assault tactics, and equipment. Atoll warfare was entirely new to our forces, and we had never assaulted a heavily fortified island before. The amtracs were previously untried for assault, and saved numerous lives by carrying the Marines across the 700-foot reef, saving them wading ashore fully exposed. Lesson learned, and future amtracs would be more powerful, better armored, and fitted with a rear ramp.

Perhaps feint attacks on the south and west sides of Betio would have spread out the defenders more, allowing the actual attack on the north beaches to face less firepower. A lull in bombardment also allowed the enemy to move men from the south beaches to the north. Sending more tanks in sooner could also have saved lives, but the dual-purpose guns needed to be taken out first. A faster means of getting the wounded to ship sickbays would have saved many, possibly by using small, shallow-draft, fast "medevac" (medical evacuation) boats.

Harold Goldberg

In essence, this question deals with the development of amphibious warfare. (Thanks to Colonel Joe Alexander, USMC [Ret.], for his outstanding work on this subject!) While we can look back and blame the American military for not studying the tides or building sufficient landing vehicles or knowing how to cross a reef, it seems to me that the entire concept of landing on a small island surrounded by a reef that opened into a lagoon that was covered by Japanese fire was a lot more difficult than Americans remember these days. This type of landing was a fairly new concept in naval/marine tactical assaults. The U.S. military had to learn how to conduct these operations someplace, and Tarawa turned out to be one of the islands where some of those lessons were learned—the hard way.

Without forgetting the mistakes at Tarawa, I am impressed by how quickly the Navy and Marines applied the new lessons. If we examine the landing at Saipan in June 1944, we see that the United States sent an underwater demolition team to check for mines, to measure the reef, and to check the tides. In addition, the Navy arrived with improved and sufficient landing vehicles; while some did not make it over the reef due to Japanese fire, the Saipan landing put 20,000 Marines on the beach on the first day. Even so, and despite improved landing vehicles at Saipan, the current combined with intense Japanese fire forced a lot of Marines to land on the wrong beach.

In retrospect, it is difficult to imagine crossing a reef into a lagoon when the enemy has all of its guns aimed on that spot. Under the best of circumstances there will be high casualties. With hindsight, we can imagine a lower casualty rate at Tarawa, but in no case should we assume an easy victory. Wherever the Japanese defended an island—Tarawa, Saipan, Peleliu, Iwo Jima, Okinawa—the fighting was brutal and costly.

To return to the issue of amphibious warfare: even if you hit the exact beach you have been prepped for, naval bombardment has probably destroyed all of your landmarks. Your map is useless. What is left of your company or platoon is scattered all over the beach. Your weapon may not be working. Your officer may be a casualty. You have a lot of issues to overcome. This is one case (amphibious warfare) where I think it may be easier to criticize than to stand back in amazement that there were so many Marines and others willing to do that job. We should honor those brave men who sacrificed themselves on that island while at the same time remembering that the war in the Pacific was a "learn as you go" experience.

The Turkey Shoot

After Tarawa, U.S. forces continued to dismantle Japan's empire, grabbing Kwajalein in the Marshalls and raiding Truk, home of the Combined Fleet for nearly two years. At the same time, MacArthur was leapfrogging past enemy strongpoints on his way back to the Philippines.

Then in June 1944, the Central Pacific Force, led by Vice Admiral Raymond Spruance, targeted the Marianas chain, about 1,300 miles southeast of Japan. The airfields on Saipan, Tinian, and Guam would allow B-29s to go all the way to Japan. Spruance's firepower came from Vice Admiral Marc Mitscher's Task Force 58, which boasted fifteen fast attack carriers, seven battleships, more than twenty cruisers, troop transports, and nearly nine hundred planes.

To stop the Americans, the badly outnumbered First Mobile Fleet, commanded by Admiral Jisaburo Ozawa, planned to employ carrier aircraft and land-based planes. Ozawa's battle plan literally went down in flames. On June 19, in the battle known in U.S. Navy circles as the "Great Marianas Turkey Shoot," the Americans knocked down nearly four hundred Japanese planes at a cost of just twenty-nine of their own aircraft.[8] Japan lost 445 pilots and aircrewmen.[9]

U.S. subs also dispatched two carriers, *Taiho* and *Shokaku*. Then, after search planes finally tracked down Ozawa's fleet, Mitscher delivered his own late afternoon attack from three hundred miles away, sinking the carrier *Hiyo* and damaging three others, *Chiyoda*, *Junyo*, and *Zuikaku*.

As his pilots returned in the dark, Mitscher ordered his carriers to turn on their lights to assist with a precarious nighttime landing. It was a harrowing experience, but 140 of the 226 attack planes returned safely.[10]

Spruance would eventually receive criticism for his conservative approach at Philippine Sea, for covering the Saipan landings at the expense of pursuing Ozawa's carriers.[11] However, he had decided in advance that his main mission was to protect amphibious shipping off Saipan.[12]

Q: *Could the Japanese have employed a better strategy at Philippine Sea?*

Jon Parshall

Japanese carrier operations during the Marianas battle were surprisingly sophisticated, actually. Their scouting was good. They had functionalized their carriers, using specific decks for attack, or scouting, or fleet defense. Their aircraft were now using more modern Finger-4 formations. But the problem wasn't one of strategy, or doctrine, or tactics anymore. The problem was simply that the material disparity, both qualitative and quantitative, between the two combatants had become insuperable for the Japanese. They could have fought at Philippine Sea with biplanes and not done appreciably worse than they actually did with Zeros and Jills and Judys. Given what they were up against, they were headed for a massacre under almost any conceivable set of circumstances.

Anthony P. Tully

When the circumstances are carefully reviewed, it is difficult to escape the conclusion that CinC Combined Fleet Koga, and then Toyoda after him, had with the A-Go plan devised the best possible use of the resources at Japan's disposal at Saipan. Ozawa carried out the plan skillfully and with perseverance, but the fact is the incomplete training and skill level of his pilots (and those of land-based air as well) were simply not up to the task. It is hard to see where the strategy itself, rather than the crippling lack of crack pilots, impaired the Mobile Fleet's chances for success. A related factor is also logistical: the two

Contrails from fighter aircraft mark the sky over Task Force 58 (TF-58) during the famous Marianas Turkey Shoot, Battle of the Philippine Sea.

most crucial ship losses—big fleet carriers *Taiho* and *Shokaku*—were in part due to the lack of adequate destroyers and antisubmarine protection. Again, the strategy itself was not the culprit. When the moment of testing came, the "shuttling" of aircraft from carriers to island bases was accomplished, but the problem was so few survived to be shuttled, instead being shot down in droves over Task Force 58. Other Japanese aircraft fell victim to raids on their bases.

Jim Hallas

Given the disparity in forces, it is difficult to imagine how Admiral Jisaburo Ozawa could have prevailed at the Battle of the Philippine Sea. The Japanese mustered three heavy carriers, six light carriers, and five battleships against Ray Spruance's seven heavy carriers, eight light carriers, and seven battleships. The quality of Japanese pilots had dramatically declined, Japanese aircraft were outclassed, and the newly introduced U.S. battle-line tactic—essentially an antiaircraft barrier protecting the carriers—all made a Japanese victory unlikely.

Considering these obstacles, the Japanese strategy in the Battle of the Philippine Sea was not ill-conceived. Admiral Ozawa planned to use the Japanese airfields at Guam, Yap, and Rota as unsinkable carriers. Before the fleets clashed, Japanese land-based planes would pound the American ships until the odds were more even. Ozawa would then take advantage of the longer range of his carrier planes to stand off and destroy the American carriers. His carrier pilots would attack, land at Guam to refuel and rearm, then attack again as they returned to their home carriers.

Unfortunately for the Japanese, Spruance was well aware of the dangers posed by the island airfields. His carrier planes swarmed the enemy fields, cratering runways and destroying land-based aircraft staging in from other bases. Japanese pilots that did manage to fly out of the island airfields were shot down while carrying out ineffectual attacks on Spruance's ships. Ozawa, assured by Vice Admiral Kakuji Kakuta, commander of the base air force of the Marianas located on Tinian, that the U.S. force had suffered heavy losses, launched his carrier planes on June 19 only to find Spruance's power intact. Japanese naval air power was destroyed in the clashes that followed.

Q: *Could Spruance have finished off the Japanese carriers at Philippine Sea? How might that have impacted the Battle of Leyte Gulf in October?*

Harold Goldberg

Admiral Spruance was in overall command of the Marianas operation while Admiral Mitscher commanded Task Force 58, made up of the fast carriers that had wreaked havoc on Japanese bases throughout the Pacific prior to this engagement. Spruance had a lot on his mind as the battle for Saipan got under way: he had to deal with information that the Japanese navy was moving from the Philippines toward the Marianas, counter that threat, and protect his invasion force of Marines and soldiers at the same time. The latter obligation was his primary responsibility, while Mitscher viewed the situation from a narrower perspective and saw an opportunity to destroy the Japanese fleet. To do so, Mitscher wanted to sail full speed toward the Japanese navy, but Spruance worried that the Japanese fleet might have been divided into two or more smaller task forces that would try to outflank the Americans. His experience

A Japanese aircraft carrier takes evasive action while under attack by Task Force 58 aircraft, Battle of the Philippine Sea, June 19–20, 1944.

at Midway certainly suggested that this was a real possibility. In this regard, Spruance was concerned that a full-speed run at the Japanese ships could leave the Marianas unprotected and vulnerable to a flanking action. Further, all of the noncombat ships in his fleet had been moved to the east of Saipan, and again he feared the consequences of sailing too far from these vulnerable ships.

In the end, both Mitscher and Spruance were correct in the following ways. We now know that the Japanese had not divided their fleet into smaller task forces that might sneak behind the American forces or threaten Saipan. So Mitscher's demand to go aggressively after the Japanese fleet might have been successful if implemented. But given the information available at the time, and given Spruance's greater responsibility in regard to the entire invasion force and the men on Saipan, the commanding admiral made the appropriate decision. He protected the Marines and soldiers in the invasion force; he established a defensive perimeter around his aircraft carriers so that Japanese planes had to fly past American planes and through antiaircraft fire before they could approach the U.S. fleet. In the end, Spruance, cautious as he was, made the right decision. Japanese planes, whether flying from their aircraft carriers or land bases in the region, were decimated by American planes.

Mitscher should have celebrated this overwhelming victory with Spruance. The infamous Great Marianas Turkey Shoot was a credit to both fine admirals, Spruance and Mitscher.

Anthony P. Tully

In my opinion, judging from the Japanese motives and options, even if Philippine Sea had resulted in the sinking of more carriers, let us say *Zuikaku* and *Chiyoda*, for example, the Leyte Gulf battle would still have taken place. In fact, it would have followed almost exactly the same pattern—arguably even more so. Why? Because the decoy to the north (and Vice Admiral Shoji Nishimura's to the south) would have been more necessary than ever. And if the Japanese had lost more at Philippine Sea, it would not have hindered this assignment. Because the flight decks to "lure" Task Force 38 would still have been available—Carrier Division 1—the *Amagi* and *Unryu*, possibly even *Katsuragi*, simply would have been sent out with whatever planes and ships remained of the Mobile Fleet to perform the same sacrifice. These full-size carriers were operational (for diversion purposes certainly so) in October 1944. The only reason they didn't go to Cape Engano with *Zuikaku* is they had no ready aircraft that hadn't already been contributed. In fact, Halsey had intelligence that at first suggested *Amagi* and *Unryu* were with Ozawa. Therefore, if Ozawa had lost more at Philippine Sea, it can be assumed that Carrier Division 1 would have been given the decoy assignment and the available aircraft loaded on board them to do as best they could.

From the foregoing, there is no real scenario where the Japanese have no carriers to send to Leyte. The real question was whether they had any trained air groups. Without the four carriers of Ozawa's Third Fleet, Carrier Division 1 would have had to very dubiously "fake it" as far as a carrier force stands. Yet they probably would have attempted to "fake it." The key to the Japanese dilemma at Leyte Gulf is the squandering of the carefully conserved air strength hurled into—and lost—in the grueling air battle of Formosa from October 10 to 15, 1944, on the very eve of the Battle of Leyte Gulf, when the Sho-Plan would most need and have cause to mourn them.

Saipan

As the Turkey Shoot unfolded in the skies above the Marianas, U.S. ground forces captured Tinian and Guam. However, the real bloodbath occurred at Saipan, where nearly 20,000 Marines of the 2nd and 4th Divisions landed on June 15. Other Marine units and the Army's 27th Infantry Division soon joined the fight.

It took weeks to overcome Saipan's defenders and its mountainous terrain. Along the way, Marine General Holland Smith relieved the 27th Division's commander, Army General Ralph C. Smith.

The United States suffered nearly 16,000 casualties; almost 29,000 Japanese died.[13]

Q: *Was the pre-invasion bombardment sufficient at Saipan? What if Holland Smith had not relieved Ralph Smith?*

Jim Hallas

The shortcomings of the pre-invasion bombardment at Saipan cost the lives of many Marines. Early bombardment was conducted by the fast battleships, which were not properly trained in shore-bombardment techniques. Fear of mines kept the battleships well offshore—in excess of ten thousand yards—which affected the accuracy and effectiveness of the shelling. Lack of training among the spotter plane pilots detracted further from the effectiveness of the bombardment. Some targets of little or no military value—such as the big sugar mill in the 4th Marine Division landing area—were repeatedly shelled, while artillery positions on the ridges behind the beaches went unnoticed and unmolested.

The situation improved somewhat after the older, slow battleships arrived on June 14, the day before the landing. Their crews were specially trained in the meticulous techniques required for effective shore bombardment. They spent the day methodically working over the enemy defenses. It was not enough.

The price of this failure was paid during the first forty-eight hours of the invasion (and was aggravated by a directive that targets farther than one thousand yards inland were to be attacked only by aircraft until H-hour). Enemy artillery and mortars that had escaped the pre-invasion bombardment rained shells down on the landing craft and the Marine beachhead, inflicting thousands of casualties, disorganizing the assault, and raising havoc with the American timetable. It took days to seize objectives that were expected to fall within hours. An extra day or two of careful fire by the well-trained crews of the slow battleships could have knocked out more of those guns, saved many Marine lives, and allowed the push out of the beachhead to proceed more rapidly.

Saipan was also the scene of the most infamous inter-service blowup of the war when Marine Lieutenant General Holland M. Smith relieved the commander of the 27th Infantry Division, Major General Ralph Smith. Holland Smith felt the 27th was second-rate; his solution was to replace its commander. Partisans have been arguing about the wisdom and fairness of that decision ever since.

There has been an effort in recent years to rehabilitate the reputation of the 27th Infantry Division, but a close reading of the record—including the division's own unit history, which was written by one of its foremost apologists—indicates that the division did not perform very well on Saipan. Some of the failings were not entirely the division's fault: for instance, the ground the GIs were forced to attack over at Death Valley in central Saipan was some of the most heavily defended terrain on the island. In defense of Ralph Smith and his GIs, it also seems evident that Holland M. Smith and his staff did not appreciate the extent of the enemy defenses facing the division and offered more criticism than support.

However, a reading of the 27th Division's history and action reports also reveals a pattern of less excusable failures, inefficiency, and an overall lack of aggressiveness. Some of these shortcomings appeared on the regimental staff level, but there were also serious deficiencies displayed on the lower level by more junior officers who could not seem to launch attacks on time, captured ground only to give it up, and allowed their men to be stalled by even minor resistance. It is significant that the division lost four battalion commanders and one regimental commander killed or wounded on Saipan, most of them when they ventured forward in an effort to keep the assault moving.

Despite the division's failings, Holland Smith—an impatient personality at best—erred in relieving Ralph Smith. The Army general was a competent and courageous officer who risked his life at the front lines while trying to sort out the problems with his division. Holland Smith did not allow him reasonable time to resolve those problems.

Ironically, nothing was gained by Smith's relief. The division did not perform appreciably better for either of Ralph Smith's two successors. In the end, the relief of an Army general by a Marine only impaired inter-service cooperation in the Pacific theater and led to bitter feelings between the two services—a bitterness that continued into the postwar years.

Don Allen

General Holland Smith relieved General Ralph C. Smith because the 27th Division was not able to keep up with the advancement of its two flanking Marine divisions, the 2nd and the 4th. He had earlier reservations about R. C. Smith's command abilities at Makin during Operation Galvanic in the Gilbert Islands. Had he not relieved him, it is very likely that the Japanese would have taken advantage of the stall in the center and focused counterattacks there, risking the integrity of the entire line.

Because of the Japanese ability to construct extremely strong bombproof shelters, pre-invasion bombing never seemed sufficient throughout the Pacific campaign. Plunging naval bombardment may have proven more effective, if the underground shelters could be pinpointed, but that was not possible. In the question of aerial bombardment, however, more is always better, at least in defeating the enemy's morale.

Harold Goldberg

Following Tarawa, the Navy brought a lot of firepower to Saipan. Ultimately, it did not make much difference. Japanese defenses were placed behind and on the back sides of hills, so that naval shells that hit the front of a hill or passed over the top had little impact on enemy emplacements. Although the sound of the naval bombardment rattled the nerves of Japanese soldiers, the actual military impact was necessarily limited. Despite the ultimate ineffectiveness of the pre-invasion bombardment, it seems remarkable that the Navy could project as much power as it did sailing so far from any American base. Unable to seek cover and isolated from supply depots, the Navy did an amazing job logistically in the middle of the Pacific. The ineffectiveness of the bombardment was not the fault of Admiral Nimitz in Hawaii or Admiral Spruance, who was in overall command of the fleet.

In fact, the Navy put together an excellent team with Admirals Spruance, Turner, and Mitscher, and they performed very well indeed throughout this engagement. The problem with the team was the relationship between Marine General Holland Smith and Army General Ralph Smith. Holland Smith planned a three-day invasion/victory, and he was angered by the Japanese resistance and high Marine casualties. Following the tougher-than-expected landing and the threat posed by the Japanese fleet approaching the Marianas from the Philippines, the Navy had to put the reserve force—the 27th Army Division—on the island between the 2nd and 4th Marine Divisions.

Despite an ever-growing American presence on the island, the Japanese fought on, attacking at night and defending every sugarcane field and every cave. As American casualties grew in number, Holland Smith's frustration and anger increased simultaneously.

Holland Smith gave the Army two difficult missions. The first one involved taking Nafutan Point in the south, a fortified rock formation that was hard to approach without sufficient armor. At the same time, Smith removed Army companies for action elsewhere, leaving the assault troops spread very thin and without the firepower they needed. The second assignment was an attack through the mountainous center of the island—the so-called Death Valley. The Army made progress but far slower than Smith wanted. As a result of his frustration, Holland Smith fired Ralph Smith in the middle of the ongoing battle. Shortly before being relieved of command, Ralph Smith came up with a new plan for assaulting Death Valley. Nevertheless, he was unceremoniously removed from the island and did not get to see his Army replacement put his—Ralph Smith's—plan into effect.

The evidence does not support Holland Smith's charges against his Army counterpart. Holland had the authority—in terms of the chain of command—to fire Ralph, but the action was not justified by the military situation. Ralph Smith was not the reason why the invasion was going slowly. American casualties could not be reduced as long as Japanese soldiers fought and sacrificed to the last man.

Having interviewed several hundred Marine and Army veterans of this battle, I can say that the bravery, dedication, and tenacity of the American forces were incredible. At the same time, the Japanese fought to the last soldier, taking and inflicting high casualties as they fell. Often overlooked due to D-day in Normandy nine days earlier and the iconic flag-raising photograph on Mount Suribachi on Iwo Jima about seven months later, the Battle of

Saipan was one of the crucial engagements in the Pacific War. As a result of this battle, Hideki Tojo resigned as premier of Japan, and Japan lost a crucial defensive possession. In addition, the United States took the nearby island of Tinian, improved the airstrips, and placed its new B-29 Super Fortress within striking distance of the Japanese Home Islands. A year after Tinian's fall, the Enola Gay left the island for a bombing mission over Hiroshima.

Surigao

Four months after Saipan, Douglas MacArthur made good on his promise to return to the Philippines. On October 20, 1944, the U.S. Sixth Army stormed ashore on Leyte in the central Philippines, with the assistance of Admiral Halsey's mammoth U.S. Third Fleet.

Just days earlier, the commander of the Combined Fleet, Admiral Toyoda, launched Sho-Go, a desperate attempt to bring about a climactic showdown with the Americans.[14] As with past Japanese plans, Sho-Go was intricate, involving multiple units, including a Center Force led by Vice Admiral Takeo Kurita, with battleships *Yamato* and *Musashi*. Kurita hoped to slip into Leyte Gulf to destroy the U.S. invasion fleet while Vice Admiral Ozawa's Northern Force, with four barren aircraft carriers, lured Halsey's carriers north, away from Leyte. Another group, the Southern Force under Vice Admiral Shoji Nishimura, would move through Surigao Strait and enter Leyte Gulf from the south.[15]

U.S. subs *Darter* and *Dace* immediately disrupted the plan on October 23 by sinking two of Kurita's cruisers in the western Philippines. That same day, bombers from the Third Fleet damaged *Yamato* and sank *Musashi*. Kurita reversed course, temporarily delaying his transit of the San Bernardino Strait into Leyte Gulf.[16]

Meanwhile, Nishimura's Southern Force advanced through Surigao Strait on October 24–25, colliding with Vice Admiral Jesse Oldendorf's collection of surface vessels, part of Vice Admiral Thomas Kinkaid's Seventh Fleet. Oldendorf had six battleships, five of which happened to be at Pearl Harbor on December 7, 1941.

As Nishimura moved forward, Oldendorf's warships crossed the Japanese "T," allowing the Americans to fire broadsides while the Japanese returned fire only with their forward guns. The one-sided battle cost Japan two battleships,

Fuso and *Yamashiro*, and the lives of nearly four thousand men. Nimitz would later describe the U.S. firepower at Surigao as perhaps the most devastating naval gunfire in the history of warfare.[17]

Q: *In the wake of the U.S. Navy's success at Surigao Strait, what lessons had the Americans learned from earlier surface battles around Guadalcanal?*

Anthony P. Tully

There were some key ones. First and foremost, the importance of good advance scouting and accurate information about the approaching enemy before actual engagement was a hard lesson of the earlier night actions. Oldendorf in particular took pains to deploy his PT boats in a way where they could serve as advance warning posts. (It is one of the ironies of the battle in Surigao Strait that the PT skippers' enthusiasm often caused them to attack first in defiance of instructions, and lose the best opportunity in so doing to send word and strength reports to Oldendorf.) A related point, quick and clear identification, was also apprehended. In the battles around Guadalcanal and the Solomons, delays in identifying a sighting as clearly enemy or friendly cost hundreds of lives. At Leyte, this was much better managed, though serious mistakes still occurred. USS *Denver* fired on USS *Albert W. Grant* at Surigao despite some indications the target might be friendly, a blunder and friendly-fire incident that still stirs controversy. In general, though, USN forces at Leyte were clearly deployed, fully aware of each other's positions, and mindful of any overlap. Another lesson of interest was related to damage control—at Savo Island especially, ship-based floatplanes had been set afire on their catapults, providing ready markers for enemy gunfire. At Leyte, Oldendorf made sure his cruiser's aircraft were shot off and landed temporarily ashore until the battle was over to avoid this danger. There were many other such insights from the night actions.

The category of lessons that "perhaps should have been learned but were neglected" includes the importance of inter-ship communication and assignment of targets to avoid over-concentration on a particular target. However, at Leyte this continued to be somewhat neglected—the most notorious case being the concentration of air attacks on IJN's *Musashi* to the near exclusion of major damage to the rest of Kurita's powerful fleet while in the Sibuyan

Vice Admiral Jisaburo Ozawa. His carriers were used as bait at Leyte Gulf.

Sea. Finally, the most notorious lesson not learned and often persisting to the present is how divided chains of command without a unifying head too often simply fail to intercommunicate or even share perspectives and intentions. For example, in its immediate context, Halsey's decision to head north after the Japanese carriers made sense within directives given to him by Nimitz, but made little sense within the overall strategic objective of first priority being to ensure that the Leyte beachhead remain unmolested and successful with adequate air cover. This latter view was the understanding of MacArthur's headquarters, not least because MacArthur's HQ had agreed to the acceleration of the invasion timetable on the assurance that carrier air cover would be available until land-based airfields could take over.

Kurita's Opportunity

Nishimura's Southern Force was no longer a factor. However, late in the day on October 24, U.S. search planes uncovered Ozawa's decoy carriers some

Vice Admiral Takeo Kurita, who pummeled Taffy 3 at Leyte Gulf.

three hundred miles north of San Bernardino Strait. Acting under the belief that Kurita's Center Force had retreated, Halsey pursued Ozawa with carriers and fast battleships, leaving San Bernardino Strait wide open.

Kurita took advantage, pushing through San Bernardino Strait and arriving off Samar, near Leyte Gulf, at daybreak on October 25. All that stood between his powerful capital ships and U.S. landing craft was a small group of Seventh Fleet ships, Taffy 3, namely six escort carriers, three destroyers, and four destroyer escorts under Rear Admiral Clifton Sprague. Sprague quickly tossed up a smokescreen, attacked with destroyers and destroyer escorts, launched planes from jeep carriers, and hid his ships in a rain squall.[18]

Kurita's big guns ravaged Taffy 3. Down went the jeep carrier *Gambier Bay* and the destroyer *Johnston*. After pummeling Sprague for more than two hours, and maneuvering off Samar for another three hours, Kurita elected to retire, fearing a return of Halsey's carriers. It was, to say the least, a surprise ending for the Battle of Leyte Gulf.[19]

Q: *What if Kurita had continued his advance against the U.S. beachhead at Leyte Gulf?*

Jon Parshall

He sinks a bunch of jeep carriers, eventually some empty transports, and then is himself sunk in the neighborhood, possibly by Kinkaid, or almost certainly by Halsey's returning forces. Either way, it's an empty gesture. People seem to forget, too, that those poor "defenseless" jeep carriers actually put a few hundred aircraft of their own into the air, and they succeeded in sinking several Japanese warships all on their own. The further the Japanese push, the more Taffys and escorts they're going to have to fight, and the more casualties they're going to take, even before the U.S. heavy units make their appearance.

Anthony P. Tully

Kurita's decision to turn back, to give up the drive, is one of the most controversial and mysterious in naval history. Many interpretations have been offered, so I will confine myself to one that perhaps is not invoked often—namely, the logistics of Kurita's attrition as actually experienced, past and future. While there can be little doubt that the Japanese fleet would have been wiped out off Leyte *if* Halsey had remained in position, or even if he had left a task group and some battleships behind, it is less obvious what would have happened had Kurita pressed on into Leyte Gulf after 10:00 a.m. It is the writer's opinion that the rest of Kurita's force would have gone on to suffer severe losses, and without compensating sinkings in Leyte Gulf or elsewhere to show for it.

Consider: in the running chase of Taffy 3's escort carriers, in the space of ninety minutes from 8:50 to 10:20 a.m., Kurita had had three heavy cruisers—*Chokai*, *Chikuma*, and *Suzuya*—mortally crippled, and one, *Kumano*, already was out of the battle. His last two heavy cruisers, *Tone* and *Haguro*, had absorbed mounting damage by 2:00 p.m. One of his battleships, *Kongo*, had received severe structural damage to its hull and props by near misses. Both of Kurita's light cruisers—*Yahagi* and *Noshiro*—had suffered bomb damage that reduced their maximum speed. One destroyer, *Hayashimo*, had been crippled by near misses alone. While it is true that Kurita's battleships were more or less shipshape by the time of his final turn away at 12:35 p.m., this was a fact that could have changed at any moment. *Kongo* would founder after being hit by two submarine torpedoes on the return to Japan, so it is easy to see where just a couple of solid torpedo hits on either *Kongo* or *Haruna* could have put them in serious jeopardy, and the stouter *Nagato* and *Yamato* could take only so much more themselves. Kurita's destroyers were his strongest remaining strength other than the battlewagons, but had expended many torpedoes.

Romantic notions of the Battle off Samar, the centermost action of the Battle of Leyte Gulf, tend to paint the picture that Kurita was on the verge of wiping out everything afloat when he broke off. In fact, the dogged and heroic stand of Taffy 3 had arguably fought him to a standstill, to the point of material exhaustion. If Kurita had pressed on into Leyte Gulf, he would have been further and more closely engaged by aircraft from Taffy 2 and Taffy 1 as well, with enough range now to prevent multiple re-arm and refuel strikes. If anything like the morning's attrition rate held, it would not have taken much to finish off cruiser *Tone*, which by dusk was down by the stern with considerable flooding and, as noted above, the smaller battleships could have been brought down by torpedoes if they connected (which fortunately for the Japanese that day, none did, or failed to explode if they did so). It is possible to go further along such scenarios, but that overdoes the point. Kurita's morning attrition rate in the Battle off Samar did not bode well for the survival of all but a handful of his ships by dusk, and that is leaving out any arrival by Halsey's forces. With no air cover, even his mighty force would have had to simply steam on and "take it," each minute and hour of hits grinding it down more. Expending the fleet was arguably what Kurita was expected to do, but this attrition rate may have suggested to him that he simply did not have the time left to go after the transports before being completely defeated. Later statements made clear that Kurita was somewhat "warship tunnel vision" in outlook and, faced with this situation, appears to have simply balked at expending his proud warships for transports. This may be the reason he sought so strongly to go after any reported enemy task force instead and spent critical hours groping around the ocean off Samar.

Jim Hallas

The Japanese naval force sent to wipe out the American beachhead at Leyte was on a suicide mission and Admiral Takeo Kurita knew it. Long after the war, he confessed that he broke off the attack toward the beachhead and the American transports because he had come to the conclusion that the war was lost and he did not wish to squander the lives of his men for no reason. Whether that was truly his motivation or whether he simply lost his nerve in the face of the valiant defense put up by the U.S. escort carriers and destroyers of Taffy 3 is of little consequence. Even had he forged on to the beachhead, he would have achieved little.

The reason, quite simply, was that he arrived too late. The U.S. landings had begun five days earlier. Four U.S infantry divisions were already ashore, along with tanks and enough supplies to last a month. The initial objectives had been seized. Many of the vulnerable cargo ships had already unloaded and left the area. The remaining transports and supply vessels would have scattered as the Japanese closed on the beachhead. Some may have been lost, but the majority would have escaped. Granted, it would have been a bloody fight as Kurita's force battled the American escort carriers and destroyers to get to the beachhead. But as all this was going on, powerful U.S. forces in the area would have been closing in. Kurita would eventually have been annihilated. As it was, he escaped—but just barely.

C. Peter Chen

Off Samar, as Kurita's ships attacked Taffy 3, he had no idea whether Ozawa's decoy fleet had been successful. In the heat of battle, he received a message from Mikawa's fleet noting that an American task force was located thirty nautical miles to the northeast, and Kurita had to make a choice under stress: Would he push on and eventually hit the transports, or would he turn around to look for the American warships? Following his neo-Mahanian thinking, he chose the latter; as history would tell, the report of the American task force was done in mistake, thus Kurita was said to have missed his chance to destroy a transport fleet critical to the American efforts in the Philippines.

What if Kurita had chosen to ignore the report of the American warships and pushed on? Given the composition of Taffy 3, it was unlikely that the Americans would have been able to stop Kurita's surface fleet, and the Japanese warships very likely would have destroyed the bulk of American transports ahead. This would have seriously hampered American efforts on the ground, possibly setting back progress at the Philippines for at least several weeks. However, this would have come at a very high cost for the Japanese. As Kurita moved his ships farther south, he very likely would have met Halsey's carrier fleet, previously lured into attacking Ozawa's decoy fleet but now speeding south for Samar. With no support in the air, Kurita's fleet likely would have seen a repeat of Sibuyan Sea. In the absence of Japanese carriers, American pilots would have attacked the largest target they could find, undoubtedly Kurita's flagship *Yamato*, which would have been as helpless as her sister ship *Musashi* at Sibuyan Sea. With MacArthur's progress slowed due to the loss of transports, the conquest of the Philippines would have taken at least several weeks longer to complete, thus pushing back the schedule of major campaigns such as Iwo Jima and Okinawa. With major garrisons such as those two listed still under Japanese control, one might wonder whether the selection of targets for the atomic bomb would have changed, though that would probably have too many variables to consider to make a good "what if" scenario.

Taking the Bait

As Taffy 3 battled for its life, Halsey was at Cape Engano, nearly three hundred miles away, chasing down Ozawa. Frantic pleas from Nimitz and Kinkaid forced a frustrated Halsey to leave two carrier groups to finish off Ozawa while Vice Admiral Willis Lee's fast battleships sped south to assist Taffy 3, arriving too late to catch Kurita.

In all, Leyte Gulf cost the United States a light carrier, two escort carriers, two destroyers, a destroyer escort, and nearly three thousand men. Japanese losses were much more extensive: four carriers, three battleships, six heavy cruisers, four light cruisers, nine destroyers, and nearly ten thousand men. Ozawa had suffered mightily, but had drawn Halsey away from the invasion beaches.[20]

Q: *What if Halsey had not pursued Ozawa's carriers?*

Keith Allen

If all of TF 38 had remained in the Leyte-Samar area, Kurita would probably have been destroyed. Of course, much speculation centers on a possible

Admiral William F. Halsey. His decision to pursue Ozawa's carriers made life interesting for Taffy 3.

battle-line action, had Halsey left Lee (and maybe one carrier group) back; in that event Kurita probably would have been defeated and turned back, but not annihilated.

Halsey did not know the Japanese carriers were decoys. But there were officers under him, like Rear Admiral Gerald F. Bogan and Vice Admiral Willis Lee, who were opposed to the abandonment of San Bernardino Strait, so I don't think this is just hindsight. And I don't think it was an either/or choice; Halsey had a sufficient margin of strength over the Japanese that he could have left one carrier group and the battleships behind, and chased Ozawa with the rest.

Frank Snyder

If Halsey had remained off San Bernardino Strait with all of the ships in his three task groups of Task Force 38, *and* if Kurita had exited the Sibuyan Sea through San Bernardino Strait when he did, about midnight, the result should have been another smashing American victory, like the battle being fought at the same time in Surigao Strait. The approach of the Japanese, through a strait, should have given the Americans the sort of advantage that they exploited at Surigao Strait.

Halsey planned for the battle to be a gunnery battle between battleships and cruisers on both sides, while at Surigao Strait the most effective weapon used by the Americans was the destroyer-fired torpedo. The six battleships in Halsey's force were all of recent construction, with 16-inch guns. Unfortunately, during 1943 and 1944, these battleships had not been employed as battleships (in the classic sense) but as platforms for antiaircraft protection for aircraft carrier task groups. Some battleship skills like observing fire-distribution doctrine or spotting fall-of-shot, had not been much practiced, so an American victory in a battleship shoot-out would not have been a sure thing.

I have not seen Admiral Lee's plan for the battle, but if he had not planned to use destroyers to attack the Japanese ships as they emerged from San Bernardino Strait then the outcome off San Bernardino might not have been as successful as the battle at Surigao Strait—where the Americans had used their destroyers. The Americans also profited by the decision of one Japanese admiral to enter the strait without awaiting the arrival of another Japanese admiral.

Donald Goldstein

We have to remember that the Japanese carriers of 1944 were not the Japanese carriers of 1941. The equipment, planes, pilots, and crews were not of the same quality. Their planes were not as good in 1944 and the Japanese pilots were not as well trained. I'm not sure it really would have mattered. By 1944, the Japanese were finished.

The United States might have been better off bypassing the Philippines at any rate. I think the Americans' best bet was to bypass the Philippines, as they had done with Rabaul, and go after Formosa instead, just as Nimitz wanted to do. We could have isolated the Japanese on the Philippines, as we had done on other islands in the Pacific, and perhaps the American and Filipino casualties would have been much less.

Q: *What if the Japanese plans at Leyte Gulf had been carried out to perfection?*

Anthony P. Tully

Nishimura's mission was to draw U.S. forces south out of Kurita's way by driving into Leyte Gulf first. He expected to be destroyed, but if by some fortune he reached Dulag or Tacloban in strength and a melee ensued that Kurita then joined in an hour or so, the worst-case scenario for the Allies would have been the destruction of several valuable transports and possibly even command ships like the USS *Mt. McKinley* and USS *Wasatch*, along with many of the Seventh Fleet's destroyers and escort carriers. Though the bulk of General Walter Kreuger's Sixth Army had landed by October 25, the loss of life would not have been small, and any bombardment of the beachhead would have been disastrous. However, this best-case scenario for the Japanese could hardly avoid an even greater loss of men and ships than they suffered, for every hour Kurita remained in the Leyte area brought his fleet closer to being fully engaged by McCain's TG (Task Group) 38.1 aircraft, and possibly even TG 38.2 and Lee's battleships rushing back down from up north. It is one of the catch-22s of the Battle off Samar that if Kurita had not made his controversial decision to withdraw, it is highly likely a pincer movement would have closed on him, as Oldendorf came back north (after destroying Nishimura's Third Section and sending Admiral Kiyohide Shima's Second Striking Force into retirement) and Halsey charged south. Again, Kurita may well have destroyed the transports and badly damaged the beachhead, setting the invasion timetable back severely, but the cost would have been the bulk of his fleet, far more than he actually lost. The original Japanese plan called for Kurita to withdraw his fleet from Leyte Gulf down through Surigao Strait, but he would have had to contend with Oldendorf's surface ships and destroyers in doing so. Though Seventh Fleet had depleted considerable shell and torpedo stocks in the Battle of Surigao Strait, it seems more than likely that sufficient strength remained to Oldendorf to delay Kurita's escape by a fatal amount of time, which would have allowed TG 38.1 (whose planes were first bombing Kurita by 1:30 p.m.) to close within good attack range.

Q: *What if Halsey had commanded at Philippine Sea and Spruance at Leyte Gulf?*

Jim Hallas

Historical hindsight indicates that Admiral Ray Spruance was the right man for the Battle of Leyte Gulf while Admiral Bill Halsey was the right man for the Battle of the Philippine Sea. Of course, the fates decided otherwise: Halsey ended up at Leyte, while Spruance commanded at the latter clash.

Ironically, the two battles presented similar sets of circumstances. Each involved a naval engagement while U.S. naval forces were covering an amphibious landing. At Leyte Gulf, the highly aggressive Halsey left the landing area largely uncovered while he chased after enemy carriers. As it turned out, the carriers were a decoy. Halsey's rashness left the beachhead open to a flanking attack by the main Japanese force. An escort carrier, two destroyers, and a destroyer escort were sunk, and the landing force was placed in jeopardy.

At Saipan, by contrast, Spruance, though far outnumbering the enemy force approaching the Marianas, hung back. Fretting about possible flank attacks that never materialized and citing his responsibility to protect the Saipan beachhead, his extreme caution cost him a chance to destroy the enemy fleet. Had Halsey been in command, it is likely he would have reacted more aggressively—as he did at Leyte Gulf—and inflicted more damage on the Japanese. In the one case, he courted disaster; in the other, he might have achieved one of the most decisive naval victories of the war.

While Spruance came in for much criticism for his excess caution at the Battle of the Philippine Sea, it bears noting that: 1) protection of the beachhead was his main responsibility, and 2) by the end of the battle Japanese naval air power had been destroyed, leaving the enemy carriers largely superfluous. The victory could have been more sweeping, but a victory it was. Halsey's lack of caution at Leyte Gulf must be viewed as the more egregious error of judgment.

Summary

The year 1944 was a disastrous year for the Imperial Japanese Navy. Defeated in the final two carrier clashes of the war, the IJN essentially ceased to exist.

At Philippine Sea, the Japanese lost nearly four hundred aircraft. Four months later, they sacrificed their remaining aircraft carriers for a shot at the Leyte invasion beaches and nearly succeeded, if only for a short time.

The year 1945 would bring U.S. forces to Japan's doorstep. Without the means to keep the Americans at bay, the Japanese faced some unsavory choices: exhaust the Americans, surrender, or die.

Ten

A TERRIBLE CLIMAX: 1945

BY 1945, JAPAN'S SITUATION had become desperate. The Imperial Japanese Navy, which had ruled the seas in the opening months of the war, was now incapable of contesting any enemy advance.

While MacArthur re-conquered the Philippines, a task largely completed by the spring of 1945, Nimitz focused on a couple of islands that would facilitate the bombing of the Japanese homeland.

Iwo Jima

The first of those islands was Iwo Jima. Eight square miles of rock and ash, Iwo was part of the Volcano group of the Bonin island chain, less than seven hundred miles southeast of Tokyo. U.S. military planners wanted it as a fighter base to support bombers on their way to Japan. B-29s using Iwo Jima could deliver larger payloads, and damaged B-29s and bombers low on fuel could use it as an emergency landing field.[1]

The Americans pounded the island for more than two months, as 21,000 Japanese, led by Lieutenant General Tadamichi Kuribayashi, situated themselves inside a network of underground bunkers and caves, many on Mount Suribachi, the 556-foot eminence that dominated the south end of the island.[2]

On February 19, 1945, the Marines landed. By D-plus-4, the Stars and Stripes flew over Suribachi.

Conquering the rest of Iwo Jima would take weeks and claim over 6,800 Americans, the highest casualty rate of any engagement in the history of the U.S. Marine Corps.[3]

Iwo Jima is about halfway between the Marianas and Tokyo. The 550-foot extinct volcano, Mount Suribachi, dominates the south.

Q: *What could have been done to reduce U.S. casualties on Iwo Jima?*

Harold Goldberg

The Japanese learned the lessons of Saipan, where they had emerged from defensive emplacements too soon and attacked American lines. In that engagement, they had wasted the opportunity to make the Americans come to them and flush them out of their caves and pillboxes. By continually attacking, the Japanese exposed themselves to the full force of American firepower.

On Iwo Jima and later on Okinawa, Japanese forces dug into the island to protect themselves from the bombardment. Without bunker-busting shells, as we now call them, the Navy's chances for success were limited. A ten-day bombardment would have killed a few more defenders and destroyed a few more caves, but it would not have changed the ultimate need for Marines on the ground assaulting enemy strongholds.

The second problem with a ten-day bombardment was the danger of leaving ships sitting in the water around the island for that long. A ship that is not moving is an enemy target. We now know that the Japanese submarine fleet had been largely decimated, but we could not be certain of that early in 1945. The Navy was understandably nervous about a ten-day commitment that might have left the fleet vulnerable to attack. In this case, I would defend the Navy decision. Only Marines on the ground could flush Japanese defenders out of their defensive strongholds.

The final question is, "Was it worth it?" There remains conflicting evidence about Iwo Jima as a refueling site. We would like to think that the number of pilots saved exceeded the cost of taking the island. The number of pilots and planes saved after Iwo Jima never equaled the more than 27,000 casualties (seven thousand dead). Intensive bombing of Japanese radar installations and airstrips on Iwo Jima, without an invasion, might have been a better use of American resources.

Don Allen

If you could quantify the number of bombs it would take to save one American Marine or soldier or destroy one of the enemy, I'm sure that every objective would be carpet-bombed mercilessly for weeks. Nobody really knew how honeycombed Iwo Jima had become with tunnels and caves, which were, for the most part, impervious to shelling and bombing. More bombing on Iwo would have had more of a psychological effect than a physical one.

In retrospect, very few of the Pacific land battles were absolutely necessary. Once we had control of the sea, above and below, nearly all of the islands held by the Japanese could have been starved out by blockade. Iwo's airfields are, perhaps, one exception. They did allow damaged bombers a closer field to land at than their home bases, which undoubtedly saved lives and planes.

Okinawa

The next island coveted by the United States was Okinawa, southernmost of the Japanese Home Islands, 350 miles southeast of Kyushu. Okinawa would serve as a bomber-fighter base and a staging area for the invasion of Japan.[4]

Landings began on April 1, as more than 100,000 Japanese squared off against the newly created U.S. Tenth Army, which included two Army and two Marine divisions. Japan's commander, Lieutenant General Mitsuru Ushijima, planned to make his final stand on the southern end of the island.

After two weeks of fighting, Tenth Army commander Lieutenant General Simon Bolivar Buckner Jr. had little to show for his extensive casualty list. A massive assault on April 19 on Kakazu Ridge, a Japanese stronghold, resulted in heavy U.S. losses. Then in early May, the Japanese withdrew to new positions and launched their own costly, unsuccessful attack.[5]

Finally, on May 11, Buckner assembled 85,000 men, including the 1st and 6th Marine Divisions, and assaulted the Shuri Line, an eight-mile arc near the southern end of Okinawa. Ten days later, the Americans achieved a penetration and then broke through the Kiyamu Line, last of the Japanese defenses. Organized resistance ended on June 22, but Okinawa claimed over seven thousand Americans and approximately 110,000 Japanese.[6]

Q: *Was Buckner's decision to attack Japanese positions head-on the correct one? If Japanese resistance had not been so fierce, would the United States have been more likely to invade Japan, rather than use the atomic bomb?*

Jim Hallas

A month after the landings on Okinawa, three U.S. divisions were hammering futilely against the powerful Japanese Shuri Line on southern Okinawa. Casualties mounted. Gains were negligible. Offshore, the supporting ships came under unrelenting kamikaze attacks. Thousands of sailors were being killed or wounded as the Navy remained tied in place to support the stalemate ashore.

Here's what could have happened: In a master stroke, Tenth Army commander Lieutenant General Simon B. Buckner broke the stalemate by sending the 2nd Marine Division on an end run—an amphibious landing over the Minatoga beaches on southeastern Okinawa to the rear of the enemy fortified line. Buckner's counterpart, Thirty-second Army commander Lieutenant General Mitsuru Ushijima, had previously stationed the 24th Division and 44th Independent Mixed Brigade in the south to deal with just such a contingency, but by early May he had been compelled to strip those forces to fill losses suffered on the Shuri Line. The beaches and outlets were now only weakly defended.

Storming ashore and moving quickly north, the Marines smashed into the Japanese right flank from behind while the 7th Infantry Division attacked from the front. The combined effort tore open a hole that allowed Buckner to turn the Japanese flank. The Shuri Line collapsed. Disorganized Japanese elements continued to fight on, but by the end of May, the campaign for Okinawa was largely over.

Unfortunately for the GIs and Marines beating themselves to pieces against the Shuri Line, the end run over the Minatoga beaches never took place. There was no amphibious assault, no quick collapse of the enemy defense and no speedy victory. The reality was far grimmer. Though repeatedly urged to conduct an amphibious end run at Okinawa, General Buckner insisted on hammering away in a frontal assault on Ushijima's fortified line.

He persisted in spite of a growing number of high-ranking officers who, sickened by the meat grinder consuming American GIs and Marines, urged Buckner to break the stalemate with another landing. One of the earliest proponents was Major General Andrew D. Bruce, whose 77th Infantry Division had conducted a highly successful amphibious end run at Leyte the previous December. Even before the American assault bogged down in the drive south, Bruce had agitated for a similar assault at Okinawa.

Buckner demurred. Though a popular and capable officer, he was conservative by nature; he was not a risk taker. It is also likely that his lack of experience with amphibious assaults may have limited his understanding of the capabilities of a Marine division in carrying out a landing on hostile shores; he maintained it would be impossible to adequately supply such a landing, that the Minatoga beaches were insufficient and dominated by higher ground, and that enemy forces in the area were too powerful. A landing behind the Shuri Line would be too costly, "another Anzio, only worse," declared Buckner.

In fairness to the general, his caution had some merit early in the campaign when General Ushijima, anticipating a potential landing to his rear, kept the 24th Division and 44th Independent Mixed Brigade in the south to guard against that eventuality. But toward the end of April, Colonel John W. Guerard, deputy chief of staff for operations and plans for XXIV Corps, noticed that members of the Japanese 24th Division were showing up in the front lines. This indicated to Guerard that General Ushijima was stripping his forces to bolster the Shuri Line. Earlier opposed to an amphibious landing at Minatoga, Guerard now felt such an assault had become feasible.

Major General John R. Hodge, whose XXIV Corps was suffering heavy casualties in front of Shuri, was convinced. He urged Buckner to land Marines over the Minatoga beaches. Other proponents came forward, including senior Marines and several Army generals tired of the meat grinder at Shuri. Even Marine Corps Commandant General A. A. Vandegrift spoke in favor. Despite the growing clamor, Buckner continued to feed troops into the frontal assault, apparently believing the Japanese were near to collapse.

His obstinacy played directly into General Ushijima's strategy, which was to kill as many Americans as possible in a battle of attrition before his inevitable defeat. Ushijima was surprised but gratified when Buckner continued the frontal assaults without attempting a secondary landing. Thirty-Second Army staff believed "such a landing could be executed relatively safely and easily, and moreover, it would bring a prompt end to the fighting," according to subsequent intelligence. Colonel Hiromichi Yahara, General Ushijima's operations officer, later revealed that after the beginning of May "it became impossible to put up more than token resistance in the south."

In the end, Tenth Army did break the line, but victory came at terrible cost. U.S. combat divisions reported over 38,000 casualties of all types, including over seven thousand dead. Among the latter was General Buckner himself, killed on June 18 by an enemy shell while visiting the front lines. Naval forces supporting the operation suffered nearly five thousand dead and another five thousand wounded, with scores of ships sunk or damaged by kamikaze attacks.

Those casualties and the determined Japanese resistance had at least one effect General Ushijima could not have foreseen. They helped persuade the United States to use its most fearsome weapon—the atomic bomb—in an effort to avoid a repeat of the slaughter in an invasion of the Japanese Home Islands. It is ironic to think that a quicker, cheaper victory on Okinawa—one engineered by a successful amphibious landing to the rear of the Shuri Line—might have persuaded the United States that the use of atomic weapons was not necessary to force the Japanese government to capitulate.

Don Allen

In any situation involving extreme resistance and high casualties, there is always the option to hold your line and pound the enemy with artillery. This could have been done in several battles in the Pacific, since the enemy's supply line was cut off, and it was only a matter of time before their food, water, and ammunition were depleted. This would have lowered our casualties, but the objective was to get the job done as soon as possible, and that meant slugging it out.

General Buckner's decision to attack head-on, especially this late in the war when we knew about the Japanese steadfastness and willingness to fight to the death, may not have been the best decision. Field and naval artillery could have done a devastating job on the defenders, relentlessly whittling them down. But, again, his orders were to take the island, and that meant on schedule.

American military intelligence understood that invasion of the Home Islands would be a far greater ordeal than any of the preceding island battles. Whereas the Marianas, Iwo Jima, and Okinawa were Japanese possessions, to be aggressively defended, invasion of the Home Islands would have been, essentially, a holy war to the Japanese, in defense of the emperor. The decision to use the atomic bombs would probably not have changed.

The Atomic Bomb

A little over three months after the passing of President Franklin Delano Roosevelt, America's Manhattan Project produced history's first atomic explosion on July 15 at the Army Air Force Alamogordo Bombing Range in New Mexico. Now it was up to the new president, Harry S. Truman, to decide whether to use the terrible weapon against Japan.

On August 6, 1945, a B-29 Superfortress, the Enola Gay, *dropped an atomic bomb, Little Boy, on Hiroshima.*

Military planners estimated that an invasion of Japan could lead to as many as 1 million American casualties. To avoid that kind of bloodbath and to end the war, Truman ordered the Twentieth Air Force to drop the atomic bomb on Japan.[7]

On August 6, 1945, a B-29 Superfortress released Little Boy on Hiroshima, killing 140,000 Japanese outright. Three days later, with no surrender signal received, a second atomic weapon, Fat Man, fell on Nagasaki, this time resulting in nearly 70,000 deaths.

Finally, on August 10, the Japanese surrendered, with the stipulation that the emperor's supreme power remain in place.[8] The formal signing ceremony took place on the battleship *Missouri* at Tokyo Bay on September 2, 1945.

Q: *Could blockade and air power have forced a Japanese surrender or was the atomic bomb, or an invasion, necessary?*

Donald Goldstein

If the Americans had not used the atomic bomb, the war would have lasted another six to eight months. I don't think the United States would have lost as many men as some experts claim, and the Japanese losses would have been much higher because the Americans were bombing Japan at will. But here's the kicker. At that stage in the war, the Russians were not doing what we had expected. At that point, the Russians were looking to grab territory. We dropped the atomic bomb on August 6, the Russians entered the war against the Japanese on the seventh, we dropped the second atomic bomb on the ninth, and the war was over on the tenth. In a matter of days upon entering the war against Japan, look at what Russia got. They got the Kueriles, Sakalin, and northern Manchuria. They acquired all of the chemical stuff the Japanese had been working on. Had the war lasted just a few more months, there's no doubt in my mind that the Russians would have been at the peace table. There would have been a North Japan and a South Japan and Tokyo would have been split, much like what happened with Germany and Berlin.

We could have won the war without dropping the atomic bomb. The bomb prevented the Russians from getting fully involved in the war against Japan and helped to avoid a divided Japan. I could see a divided Japan, a North Japan along the lines of North Korea, at least until 1990 when the Berlin Wall came down.

Clayton K. S. Chun

President Harry Truman faced one of the most difficult decisions of his administration in 1945: how to defeat the Japanese and force them to surrender. Truman's end objective was the surrender of Tokyo; the Japanese did have the means to continue the war and inflict many casualties on an attacking force. The only question was the way Truman would finish the war. The president could have chosen from several strategies: American forces could expand their strategic bombardment on Japan, continue the naval blockade, invade the Japanese Home Islands, create a second front with the Soviets, or try to shock the imperial government by using the atomic bomb. Ultimately, Truman used and continued many of these options to shape the environment for a Japanese surrender. No one option forced Emperor Hirohito and the Japanese government to accept unconditional surrender. However, together these options created conditions that constrained Japanese actions and forced capitulation.

Each of the options had considerable advantages and disadvantages. With the capture of the Marianas, the Army Air Forces finally had bases, logistical support, and sufficient support to unleash massive B-29 raids over Japan. Using tactics such as low-level incendiary night attacks, B-29 crews destroyed Japanese military, industrial, and other targets with impunity. Aircrews literally destroyed cities and the means for Japan to continue the war. Similarly, the U.S. Navy had continued to pursue its War Plan Orange, developed throughout the interwar period, to strangle Japan and cut it off from its sources of raw materials and food supplies. The Navy's submarine campaign had savaged Japan's merchant marine fleet and oceanic transportation around Japan. Both campaigns had succeeded in drastically reducing food stocks and raw materials to fuel Japan's industry. Both strategies worked and would eventually force surrender, but it would take an unknown time to starve and bomb Japan into submission.

Truman, under considerable political and time pressure to end the war, faced the prospect of approving an invasion of Japan under Operation Downfall, potentially ceding control of Asian territory to the Soviet Union, and using the atomic bomb, a weapon untested in combat. Issues such as war weariness at home, mounting casualties in the Pacific, difficulty funding the war, an unknown future for Japan, and the prospect of a bloody invasion of Kyushu and Honshu dogged Washington.

National and military leaders did not know the impact of using nuclear weapons. Scientists had tested a plutonium weapon in July 1945 in New Mexico, but not on a military target. Discussion ruled out the use of either the uranium or the plutonium weapon for a demonstration due to the fear of a weapon failure that might strengthen Japanese resolve. Similarly, Pacific commanders had already witnessed tens of thousands of casualties from incendiary night bombings of Tokyo. Substituting a fleet of B-29s armed with incendiaries for one carrying a single atomic weapon did not seem hard to imagine. Still, a single atomic bomb's destructive power was not lost on Truman. He still had to wrestle with the decision to use invasion, continue the air bombardment and naval blockade, or drop the atomic bomb. If he dropped the atomic bomb and it shocked the Japanese government into surrender, then he could call off the invasion. Truman could also have waited to produce additional nuclear weapons. Likewise, if the atomic bombs failed to convince Tokyo to capitulate, then he could approve the scheduled final assault on Kyushu in November 1945 and the follow-on Honshu operation.

Washington used a combination of approaches to convince the Japanese to surrender. Unknown enemy strength, time to complete particular strategies, political agreements with the Soviets, war weariness, bureaucratic inertia, service bias, spotty intelligence, cost, and a host of other issues guided decisions to continue several approaches simultaneously. Fortunately, the multiple approach weakened Japan and, with the use of the atomic bomb, broke Tokyo's will to resist and avoided an invasion.

Nick Sarantakes

George C. Marshall and Ernest King were extremely worried about the final defeat of Japan. Their concerns, though, were significantly different. Marshall believed that the American people were time-averse; that there was only so long that they would support the war. He thought an invasion of Japan was the fastest way to bring the conflict to an end, and the U.S. military needed to pursue the fastest course possible. King, on the other hand, was far more concerned about casualties, believing the public was averse to losses in men. He worried about time, but his bigger concern was a fear that a high number of casualties would turn the people off seeing the war through to the bitter end. He, as a result, favored a blockade.

On June 18, 1945, President Truman had a meeting with Marshall and King in the Oval Office to decide between these two options. He said he was concerned primarily about casualties, but he ended up making no decision—which as a practical matter meant the United States would do it all. Now, as we all know, Japan surrendered before the United States hit Japan with all the resources Truman had at his disposal. An invasion was not required to end the war, and a blockade, according to many, was all that had ever been necessary.

There is a good deal of historical truth in that view. As a result, in the years since World War II proponents of both air and sea power have argued among each other that the Air Force or the Navy was solely responsible for forcing Japan's surrender.

The thing about this argument is that it avoids an important diplomatic event that took place in the summer of 1945. The United States made an important qualification about the future of the monarchy. There were two different points in the summer of 1945 when leaders in both Washington and Tokyo wrestled with this issue. Essentially the United States had to back away from "unconditional surrender," which it did in a vague fashion, much to its advantage later during the occupation of Japan. If, however, American leaders had refused to make this concession, it is difficult to see how anything other than invasion and occupation would have forced Japan to accept "unconditional surrender," and it is very possible that an invasion of Japan was beyond the resources of the United States. So, in the end, diplomacy was the difference maker.

H. P. Willmott

In terms of the end of the war so much has been written, relatively little of it complimentary, about Hirohito, specifically with reference to his alleged personal responsibilities during the war. It is somewhat hard to see, however, wherein lay elements of choice and initiative for the emperor given the prevailing attitudes within the Japanese leadership, political and military. It would seem that matters where the emperor possessed real choice were few, indeed perhaps only two in number, the first being that on February 26–29, 1936, when it was on his personal initiative and orders that the army moved to put down a mutiny. There was political and military paralysis within the Japanese system as a result of this military uprising, and it was the emperor's personal intervention that ensured the uprising was suppressed. That being noted, however, the result was that political liberalism was all but killed by this episode and the country was ruled thereafter by a political-military combination that was very largely the Imperial Army's preserve. The emperor really only had the power to make decisions when there was a lack of decision on the part of the Japanese high command. In 1941, when Japan made the decision to go to war, Hirohito could ask some questions and make a few points but he did not have the power of decision at this time and on this issue. The situation was very different in August 1945 when, in the absence of a decision on the part of the government and military to surrender, Hirohito personally made the decision to end the war: he took the decision "to bear the unbearable," sent out members of his family to enforce his decision, and he broadcast to the Japanese people—though he did so without mentioning the word "surrender."

That decision invariably raises the obvious issue, the American use of atomic weapons against two major Japanese cities. After the war was over a Japanese study came to the conclusion that had the war continued into 1946 then perhaps as many as 7 million Japanese would have died, primarily as a result of malnutrition-related disease. Japan was very literally at the end of its tether by August 1945. In 1941, Japan had imported some 48 million tons of raw materials, finished items, and foodstuffs. In the first seven months of 1945 Japan's imports mustered some 7 million tons, and Japanese industry was in end-run production, while the people were only marginally above real-hunger level: it has been suggested that the Japanese population overall was living at a level about 6 percent above starvation levels. Without rice, without trawler fleets and seafood, without the means to ship, store, and distribute Manchurian foodstuffs and resources, Japan in August 1945 was little more than a disaster in the making, and it is difficult to resist the notion that it would have been forced to surrender in winter 1945–46, even if there had been no landings and no use of atomic weapons.

The point is important because, given the American planning that provided for landings in Kyushu and Honshu, the fact was that by 1945 there were very few decent military formations in the Home Islands. The 1944 and 1945 military classes were basically untrained. There were few tanks, vehicles, and radios, and in terms of firepower and mobility, the Japanese units and formations were very poorly provided. If units stood on the beaches—and the Japanese had worked out more or less where landings had to be made—they would have been shattered by the firepower that could be brought against them, and if they were held inland they would never get forward in the face of overwhelming American air power. There is always talk of Japanese plans to arm the civilian population, but in fact the Imperial Army opposed such plans: the military high command was of the view that the civilian population was not there to defend the army. It is very hard to understand the basis of assertions that the use of atomic bombs was crucial in saving Allied lives, and that somehow landings on Kyushu and Honshu would have resulted in bloodbaths. Japanese losses most certainly would have been very heavy indeed but there is no way in which the Japanese would have been able to inflict on the Allies losses that would have been four or five times those incurred in Normandy. The use of the atomic bombs was the product of three factors: the bombs were not built not to be used, there was the belief that their use would speed the end of the war, and there was the awareness of impact in terms of the Soviet Union. But at the end of the day, the use of atomic weapons against Japan in August 1945 will never be freed from a moral dimension and the decision subjected if not to censure then considerable reservations, thus:

> It is my opinion that the use of this barbarous weapon at Hiroshima and Nagasaki was of no material assistance in our war against Japan. The Japanese were already defeated and ready to surrender. . . . My own feeling was that, in being the first to use it, we had adopted an ethical standard common to the barbarians of the Dark Ages. I was not taught to make war in that fashion, and wars cannot be won by destroying women and children.[9]

And, when informed of the decision to use atomic weapons,

> I had been conscious of a feeling of depression, and so I voiced . . . my grave misgivings, first on the basis of my belief that Japan was already defeated and that dropping the bomb was completely unnecessary, and secondly because I thought that our country should avoid shocking world opinion by the use of a weapon whose employment was, I thought, no longer mandatory as a measure to save American lives. It was my belief that Japan was, at that very moment, seeking some way to surrender with a minimum loss of "face."[10]

Such observations, with their confrontation of the moral issue, cannot be lightly discounted by virtue of source: the first observation was made by Fleet Admiral William D. Leahy, then chief of staff to Presidents Roosevelt and Truman, the second by then–General of the Army Dwight Eisenhower.

Jim Hallas

The real question is not whether the war would have ended without the atom bomb—of course the war would have had to end sometime and in some way or another—the question is how long that capitulation would have taken, what form it would have taken, and what condition Japan (and the United States) would have been in when the inevitable finally occurred.

Up until the very end of the war, there were many in high U.S. military circles that believed that blockade and bombardment would be sufficient to induce a Japanese surrender. The great unknown was the time frame for this surrender: estimates ranged from only a few months to as long as several years.

It is significant that the Japanese government had begun making peace overtures in the summer of 1945. However, the idea that Japan was on the verge of collapse or had lost the will to fight is belied by the massive and continuing preparations to contest an American invasion. Even after two atomic bombs—and following the emperor's decision to sue for peace—there remained a core of hard-liners, mostly army, but also some navy, who were determined to fight on until mass annihilation. The Japanese civilian population had no influence; the people would do as they were told.

Considering that stubborn mind-set, it is likely that the Japanese would have endured a blockade for some considerable time. As for a continued bombing campaign, the experience in Europe had demonstrated that airpower alone was not sufficient to defeat an enemy. There is no reason to believe it would have been any different in an even less industrialized Japan.

Two diplomatic approaches in conjunction with a blockade and bombing campaign might have speeded the peace process: an early U.S. guarantee that the emperor would not be removed from power and a softening of the demand for unconditional surrender. The latter, which presumably would have included consent that the Japanese army would disarm itself and that Japanese war criminals would be tried by the Japanese themselves, would have been politically unacceptable in any case.

Stephen D. Regan

The United States already had air and sea command around Japan by July of 1945. Virtually no supplies were being transported into the island nation because our submarine patrol sank transport after transport attempting to enter Japanese harbors. The data clearly show that food and requisite resources unavailable in Japan were not getting to the enemy. With a very tight grip of a sea blockade, the United States found Japan totally unwilling to discuss the end to the war. If anything, the blockade heightened the motivation of both the military and the civilian population.

General Curtis LeMay's Army Air Force bombers were destroying city after city in his fire bombings. Readers of history must remember that LeMay's firebombs killed more people than the atomic bombs. The paper-and-wood houses of Japan were made of tinder, needing little to explode into flames. Much of Tokyo was destroyed, to the point that there was nothing left to bomb, and urban targets were becoming limited. Tens of thousands of military and civilian citizens died cruelly by such tactics but still the military, the government, the civilian population, and the emperor were unwilling to even discuss among themselves the possibility of capitulation.

Even after the explosion of two atomic weapons with horrendous losses at Hiroshima and Nagasaki, the military still had over 1 million highly trained men in China and several million trained or easily trained militia in the Home Islands. Civilians were being taught weapon usage with spears or anything else that could be used as a killing device. The data overwhelmingly support that the Japanese people would have gladly died to the last person standing against the Americans. Civilian suicides at Okinawa and Saipan convinced the American authorities that such suicides of men, women, and children would indeed occur if the Allies attempted invasion.

But the most convincing argument for the use of the nuclear weapons was the onslaught of suicide planes, starting at Leyte Gulf. The American losses at the hands of kamikaze pilots were steep. A teenager with minimal training on take-off and flying could sink capital ships and kill many dozens of American sailors and Marines. Japan still held many barely flyable planes for suicide attacks. The data show that dozens of miniature submarines remained unscathed, hundreds of manned torpedoes were secreted for attacks, and even flying bombs were available for bloody action against the Americans. Again, the data are overwhelming in that the Japanese could have and would have put up a hellacious defense of their homeland.

Michael Barnhart
Japan's surrender was ultimately dependent not upon a ground invasion (as the U.S. Army contended), or the effects of a blockade (as the U.S. Navy argued), or the impact of any sort of bombing campaign, incendiary or atomic (as the U.S. Army Air Forces maintained). It depended upon the ability of elements inside Japan to pry effective control of Japan's foreign and defense policy from the hands of the Imperial Army. If the army had remained in control, no amount of bombing, blockade, or invasion would have produced surrender. The emperor's intervention in the surrender decision was indeed crucial, but even he could not have pulled off a de facto coup against his own army alone. So, what did? Two things: the realization (after Hiroshima) that the army's decision to fight to the end really would mean the extermination of the Japanese people—the Americans were not going to stop short of the Potsdam Declaration's terms—and, just as crucially, the intervention of the USSR, which guaranteed that if Japan did fight to the bitter end, there would be Soviet forces on Japanese soil by that time.

Q: *What if the United States had invaded Japan?*

Nick Sarantakes
The costs of an invasion of Japan would have been horrific. It does not matter if you are looking at this development from the American or Japanese perspective. A lot more people would have died than was actually the case.

The Imperial Japanese Army had units in almost perfect positions and numbers to defend against and perhaps even defeat an invasion of Kyushu. Even if successful, this operation would have been extremely costly. Probably not the 500,000 that President Harry S. Truman claimed, but the figure would have easily been in the tens of thousands. Hundreds of thousands is not difficult to imagine either. Those losses would have come just at the time when the American public was expecting some relief from wartime mobilization following the defeat of Germany, and when the United States was beginning to hit the limits of its available manpower.

The logical implications of an invasion of Japan are also important to keep in mind. An assault on Kyushu would have been followed by operations aimed at the other three main Japanese Home Islands. World War II, as a result, could easily have gone into 1947 or 1948. If these operations had been unsuccessful or had been seen as moving slowly or being extremely costly—what seems the most likely scenario—there would have been a lot of recrimination back home. We now know that Japanese agents in Manila got ahold of MacArthur's invasion plans. They were unable to get these plans back to Japan before they were recovered, but U.S. officials could not be certain of that fact. (People in MacArthur's headquarters went out of their way to cover up this theft.) In the wake of a less-than-successful assault on Japan, news about this breach of security would have gotten back to Washington as officers in the know began covering their rear ends. The likely result would have been that Truman would have fired MacArthur six years earlier than he actually did. The general's most likely replacement would have been General Joseph W. Stillwell, the senior four-star general in the U.S. Army, who was on Okinawa at the time. The problem was that the capable and brilliant Stillwell was already dying of cancer. If the assaults on Japan had lasted longer than a few months, Stillwell would have died long before these efforts concluded. The impact on morale of going through two commanders in less than a year would not have been particularly good for the Americans.

While the Japanese army became better at fighting on the defensive and even more motivated as the war approached the Home Islands, an invasion would have been devastating for Japan. The casualties from direct combat operations and indirect results, like the collapse of transportation networks and the inability to get food to hungry people, would have been massive. Casualty numbers would have easily been in the hundreds of thousands, and a million or two is not at all difficult to imagine.

An invasion of Japan would have been bad for Asians and postwar U.S. influence in the Pacific as well. Millions—the number is no exaggeration—would have died from combat, starvation, malnutrition, and related diseases in places like China, Burma, Korea, and Singapore. The Soviet Union would also have had more time to extend its influence into northeast Asia. Power vacuums would have developed in Japanese-held territories, which would have worked against U.S. interests.

In short, in a strategic sense, an invasion of Japan would have been "a bridge too far."

Clayton K. S. Chun

If Truman had approved Operation Downfall, then it would have been the largest amphibious invasion in World War II. However, the Imperial Japanese Army had prepared extensive defenses to include ground defenses, air and naval suicide units, and a mobilized civilian population to resist the attack. The initial landing on Kyushu would have used veteran American units from the Pacific. These units had grinded north toward Japan. As these divisions moved closer to the Japanese Home Islands, casualties had increased greatly and the ferocity of combat had intensified. Iwo Jima and Okinawa were only a taste of what was waiting for Allied forces in Operation Downfall. Additionally, Allied intelligence reports continued to expand the size and scope of enemy defenses. Given the unknown but growing enemy strength, casualties would have been very high on both sides. The Allied invasion would have been difficult. Allied invasion forces on the Kanto Plain would have had to drive north to strike Tokyo, fighting in damaged urban areas. Allied forces would have required much effort to eject Japanese defenders. Additionally, American units for this operation would have had "veteran" European divisions in the assault. Unfortunately, the Army had demobilized combat veterans from these units and had largely replaced them with green troops. Could the invasion have worked? Yes. However, the Japanese would have exacted a costly attritional campaign on the Americans in an attempt to get a peaceful settlement.

Jim Hallas

The decision to begin invading the Home Islands in November 1945 stemmed from the political unacceptability of maintaining huge military forces in the field for an indefinite period while waiting for the Japanese to give up. The American public wanted an end to the war. Morale both at home and in the military itself was not indefatigable. Hitler had been defeated; now it was time to finish Japan. Better to do it now while the means was at hand.

While a two-stage invasion was planned, it was widely believed that a direct assault would result in the collapse of enemy resistance after the seizure of Kyushu. Even so, such a victory would not have been cheap. Approximately

60,000 Allied battle casualties were anticipated during the first two months alone. The physical devastation and loss of life on both sides would have been horrific.

The United States would have won in the end, but high casualties and a recalcitrant enemy would have certainly led to a harsher occupation. American anger would have made it politically necessary to remove the emperor. Japanese social structure would have collapsed and resistance to the occupation would have intensified.

Stephen D. Regan

While the estimates of American losses in an invasion have been hotly debated among historians and military experts alike, no one can deny that a million U.S. deaths and many times greater casualty levels were very likely. On the flip side was the question of civilian losses. With the training of school children and young girls to be willing to commit suicide for the emperor while stabbing, hurling grenades, or otherwise killing U.S. Marines and Army infantrymen, the potential losses of Japanese civilians were staggering. The nuclear bombs probably saved many times the number of civilians that were killed.

Finally, we must remember, although Americans are quick to forget such things, that war is a horrible, gut-wrenching, painful, and hideous murdering of men and the similar loss of our own men. Was America willing to lose a million more of its sons on the islands of Japan? Was America willing to endure the economic costs of several more years of war? Wars are easy to start but darn difficult to end. Furthermore, war is not some nice little competition between nations. We learned that lesson in the bloodbath of World War I. War is killing and being killed. Would President Truman accept the moral responsibility of allowing additional Americans to die in order to avoid use of the atom bombs?

Personally, I feel he would have committed a grave and serious ethical failure if one additional American had been sacrificed in order to allow an additional enemy to survive. Those are the ethics and morality of war that we easily overlook. Above my desk is a photo of a ditch on Bougainville in which several corpsmen are working diligently on seriously wounded Marines of the 9th Marines. One of those corpsmen is my father. Several of the Marines in that ditch never returned home. It is a gory picture, but it accurately depicts war on a personal level. War is hell. The quickest and least costly means to end war is a moral absolute. As General George Patton is reputed as saying, "You do not win wars by dying for your country. You win by making the other son of a bitch die for his." That sums it up so much better than my meager scribbling.

War Crimes

As part of the surrender package, the Allies allowed Japan to retain her emperor, although Hirohito would eventually renounce his divine status.

However, there would be no war crimes trial for Hirohito as there had been for Hideki Tojo, the minister of war, who went to the gallows in December 1948.[11] The supreme commander of the U.S. occupying forces, General MacArthur, felt that bringing Hirohito to trial would incite guerilla warfare.[12]

Q: *What if the United States had taken a harder line in reconstructing Japan, had not allowed the Japanese to retain their emperor, or had tried Hirohito for war crimes?*

Clayton K. S. Chun

Renowned military theorist B. H. Liddell Hart commented, "The object of war is a better state of peace—even if only from your own point of view." One of the most vexing problems facing President Harry Truman was how to end the war with Tokyo and create a peaceful, reconstructed Japan. For Truman, the single condition of unconditional surrender of the Axis powers was made much earlier at the Casablanca Conference in January 1943. Unfortunately, the focus of the Allied nations' leadership was not necessarily on reconstruction or how to convince the Axis powers to surrender other than through military victory. In Germany's case, Berlin surrendered at the point of a gun. Italy capitulated. Allied forces were advancing through the southwest and central Pacific, but the costly island-hopping campaigns were getting bloodier the closer American military units got to the Japanese Home Islands. Despite the use of nuclear weapons questions remained about Tokyo's surrender. If the Japanese government did not surrender, then a bloody invasion loomed.

Getting Tokyo to capitulate under unconditional surrender terms would have been difficult. Japan retained a considerable military, despite taking grievous losses. Tokyo also had a history of never surrendering. War weariness

Fleet Admiral Chester W. Nimitz signed the surrender documents for the United States. Shown with Nimitz, from left to right, are General Douglas A. MacArthur, Admiral William F. Halsey, and Rear Admiral Forrest Sherman.

among the American public and military and a staggering war cost were also issues facing Washington. Other considerations included geographic issues, American political pressures to end the war, limited communications with the Japanese government, limited intelligence information about Tokyo's intentions, the continued and stiffening resistance among the Japanese, and concern about Allied prisoners of war (POWs). Unlike in January 1943, the United States in the summer of 1945 had demonstrated its military superiority over Japan. However, Washington still needed to convince Tokyo to capitulate. The Japanese government and public's acceptance of the unconditional surrender terms depended on the condition of the future of the emperor and the imperial system. Allowing the imperial system to exist would have disarmed a hostile public and military. If the emperor could convince the Japanese military to surrender and cooperate with the Allied powers, then reconstruction and a return of Japan to the community of nations would be easier.

Allowing the Japanese people to maintain the emperor was not a decision made lightly. The American public and many in government, including in the State Department, wanted to try the emperor for war crimes. The Japanese military had fought the Pacific War on his behalf, and not making the emperor accountable for the war would create domestic and international issues concerning the softening of American commitment to end the war. Additionally, if the Truman administration hinted that it would soften surrender terms then the Japanese government might be emboldened by this move and seek even better conditions. The Japanese government might even try to re-militarize after surrender.

Convincing the Japanese public and military to accept their defeat and creating conditions for them to willingly transform their society would be a difficult task. Washington needed the cooperation of the Japanese government for reconstruction due to a lack of language and technical capability. If Washington had indicted the emperor and held him for a war crimes tribunal, then one could speculate on the future success of the reconstruction and any peacekeeping activities throughout the Home Islands. Allowing the Japanese to first surrender with the understanding of maintaining the imperial system and then trying Hirohito would also have brought American integrity under question, especially in an era where communism was expanding throughout Europe and Asia.

American and Allied military and civil officials had a very complex set of tasks to complete in Japan. They needed to rebuild society, create and expand democracy, provide security, ensure humanitarian relief (to include repatriating Allied POWs and civilians), dismantle Japan's military capability, establish new freedoms for Japanese civilians, and prosecute war criminals. If the U.S. government had implemented a stricter and harder reconstruction, then many issues may have arisen. A harsh effort would have forced an extended Japanese reconstruction over time that would have required more occupation forces and expended limited resources in a postwar era. Additionally, it would have slowed the creation of a burgeoning ally to counter the growth of communism in Asia. This would have limited Japanese civil support to American forces and tied the latter down during the upcoming Korean War. Washington also would have had to stretch humanitarian aid to the Japanese instead of making them self-sufficient. Also, the severe treatment of the Japanese by Washington could have made pre-war claims of colonial domination of Asia ring somewhat true. Finally, a harder reconstruction might have delayed the transformation of a fascist nation to a democratic one.

Jon Parshall

Many veterans ask why the Japanese never apologized to us for their actions during the war. I respond by saying, "It's nobody's fault but ours." We wrote a new constitution for them. We allowed the emperor to stay in power. We allowed the Japanese to avoid too much in the way of soul-searching about their actions during the war. Why? Because we were desperate to have a friendly, and staunchly anticommunist, partner in the area as the Cold War kicked off, and was followed shortly thereafter by the Korean conflict. There was a window of opportunity to make the Japanese revisit their war and renounce it. We squandered that time. After the status quo is reset in a top-down, paternalistic, hyper-traditional culture like Japan's, it's very, very difficult to revisit painful topics in the past. That ship has sailed.

I personally think that Hirohito was a war criminal and should have been duly tried and executed, along with a number of members of the various pre-war cabinets, members of the larger industrial concerns, and senior military officers (particularly those in positions of command in China who were responsible for atrocities such as the Rape of Nanking). We certainly could have done so—our victory over Japan was so utter, and so complete, that they would have had no ability to resist us in any meaningful sense. I also think that forcing the Japanese to revisit their actions, and not allowing them to conveniently forget about what had happened during the war, would have paid enormous diplomatic dividends in the modern era by potentially defusing some of the more nationalist, anti-Japanese rhetoric emanating from places like China. The Chinese have every reason to be suspicious of Japanese intentions, because unlike Germany, Japan has never really come clean about its actions. Conducting a Japanese version of the Nuremberg trials, and exacting retribution against those parties who were truly responsible for planning and running the war, was the least we could have done for future generations.

Anthony P. Tully

This would have been the greatest mistake of all, for the reconstruction and reviving of Japan proceeded more smoothly than that of Europe partly because there was such a common reference point in "nominal" leadership that provided a continuity from the old to the new Japan, while at the same time compromising almost none of the victor's prerogatives in the occupation and reconstruction. It is difficult to see any real value, apart from theater and a show of reprisal, in such an action. A trial like that would have been protracted and bitter in its form, and deeply alienating to an already humbled and humiliated foe. The question is further complicated by the controversial but arguable point that such a prosecution and trial would not have been directed at those truly most responsible for driving Japan into war. Indeed, it would have strayed close to scapegoating. While Emperor Hirohito had a definite say and ultimate responsibility for going to war, and enabled those who advocated it, he was also clearly (to a degree) carried into it by the militarists and the semi-"hands-off" nature of the monarchical system as it stood. The question is not whether Emperor Hirohito bore responsibility, but rather, how much, and was any lesson/punishment a trial would impose a useful or constructive course? In all probability, the answer is no. As it was, Hirohito carried the responsibility of embodying Japan's identity from militant empire-building to democratic and economic rebirth in a careful and difficult trapeze act, which provided Douglas MacArthur and the SCAP (Supreme Commander of the Allied Powers) staff the filter and buffer needed to blunt reactionary forces while fast-forwarding Japan into the postwar world. It's hard to argue with the demonstrable success of an enemy-reformed-to-staunch-ally that in fact happened, and even more difficult to see how a sensationalist trial could have in any way made that reconstruction faster or less acrimonious.

Nick Sarantakes

The arrest and trial of the emperor would have been a mistake of monumental proportions that would have had enormous and negative repercussions for decades. In December 1945, MacArthur warned Eisenhower, who had by this time replaced George C. Marshall as Army chief of staff, that a decision to arrest Hirohito would provoke an insurgency and would require 500,000 troops to contain. MacArthur had a tendency to exaggerate when it came to making statistical estimates, but this telegram is basically correct. The Imperial Army and Navy were making plans to fight an insurgency against the occupation. The trip wire would have been the arrest of the emperor. The Japanese military had plans to go into the mountains, with one of the emperor's cousins under their protection, and initiate a fight to preserve the monarchy.

Such a development would have been bad for almost everyone. Japan barely survived the winter of 1945–46. Food imports from the United States were the only thing that prevented a famine from developing. If the emperor had been arrested in early 1946 and an insurgency had started immediately afterwards, food shipments to Japan would have probably ended fairly quickly or the Americans would have had problems distributing these items to the public. Either way, a lot of Japanese civilians would have died from malnutrition and starvation.

An insurgency that winter would have caught the United States at almost the worst moment possible. The troops on occupation duty in Japan would not have been ready for the task of conducting a counterinsurgency. They were also not adequate in number. Japan is a big country, roughly the size of California, and its population then was about a third that of the United States. There would have been a need for many, many more soldiers than were stationed in Japan at that time. This new demand for military personnel would have come just as the United States was beginning to demobilize, and given the vast numbers that would have been required to fight an insurgency in Japan, voluntary enlistments would have been inadequate. Now think about what most soldiers, sailors, and Marines were thinking about at this time. They had survived a long and hard war. They wanted to get on with their lives. To be suddenly told that they were going to go fight in Japan would have had a devastating effect on morale. You would have had a lot of young men who resented being sent into combat a second time. The U.S. military that would have fought the Japanese insurgency would have been less capable than the one that defeated Germany, Italy, and Japan in the first place.

Low-intensity conflict is a slow way of fighting. Under the best of circumstances, an insurgency in Japan would have taken three to four years to resolve. The Imperial Army and Navy were making plans—just as they had during the war itself—to fight the war independent of one another. This divided approach would have made it easier for Americans to conduct successful counterinsurgency operations, but the U.S. Army has never liked fighting low-intensity conflicts. It has won many of these campaigns, but even in the nineteenth century when it was fighting against Indians, there were other missions it wanted to perform. This conflict would have been a long, slow grind with many brutalities that would have been controversial back in the United States. The American people would have had two opportunities to express their frustration at the Truman administration for the slow progress in Japan and for what some good Republican publicists would have called snatching defeat from the jaws of victory. Midterm elections always go against the party that controls the White House, even in times of war with voters usually expressing frustration at what they consider the slow progress of victory. This happened during every major conflict in the twentieth century, even World War II. The midterm election of 1946 went against Truman anyway; with an insurgency in Japan, the results would have been even worse. In this environment, Truman's upset win in the presidential election of 1948 becomes highly, highly unlikely. He barely managed to win when major factors were going in his favor. If Japan had been a basket case, Thomas E. Dewey would have become president in 1948, which had the potential to alter the American political landscape in profound ways.

The results of a U.S. victory in the Japanese insurgency would not have produced an international environment that was all that conducive to American interests in the postwar world. Intervention in Korea would have been difficult. The result would have been a unified, communist Korea that was hostile toward both the United States and Japan. The occupation of Japan itself would have lasted longer than it actually did and would have been less successful. In a best-case scenario, Japan would have been politically weaker than it was in the 1950s and 1960s. In a worst-case scenario, the Japanese would have been an angry and sullen "ally," looking for opportunities to leave the American coalition. During the 1950s, the Eisenhower administration feared that Japan was on the verge of going neutral in the Cold War, and designed its policies toward Japan to keep that from happening. A Dewey administration would have actually had to deal with that outcome. The result would have been a weaker U.S. presence in the Pacific.

There also would have been ramifications in Europe. The U.S. Army had a ready source of troops in Germany that it could have sent to Japan to fight an insurgency. Such a move, though, would have created a power vacuum in West Germany and removed a force restraining French occupation forces from imposing a harsh peace. The French were actually worse than the Soviets at this time in the economic demands they were making on the Germans—hard as that may be to believe. The result would have been a West Germany alienated from the West; the West Germans would not have voluntarily joined the Soviet bloc—no one was that stupid—but they would have been less supportive of U.S.-led efforts in the Cold War.

Jim Hallas

The Allies would have been justified to put Emperor Hirohito on trial as a war criminal. It has been argued that the emperor, due to his unique status in the power structure, actually had little or no influence on the outbreak of hostilities and the course of the war, and was merely the dupe of warmongering generals. In fact, the historical evidence indicates that Hirohito was much better informed and had more input on the conduct of the war than his apologists would like to admit.

But while the Allies would have been justified from a moral standpoint to put the emperor on trial, such a decision would have been disastrous from a practical standpoint. As far as the Japanese were concerned, the emperor *was* Japan—there could be no nation, no national identity without the emperor. The Allies needed to retain the emperor to avoid civil unrest and to ensure Japan's smooth transition from belligerent to occupied nation. (That most facile of political animals, General Douglas MacArthur, understood this practical reality and did his utmost to see that Hirohito was not charged as a war criminal.)

A harsher approach or putting the emperor on trial as a war criminal would only have completely destabilized the social and political fabric of the country, greatly complicating the task of the occupiers and making for an uncertain future. Retaining the emperor allowed a smoother transition socially and politically; the emphasis on the reconstruction of the defeated nation—a helping hand instead of punishment and reparations—allowed Japan the economic means and the dignity to recover and become a willing partner as it was reshaped into a more democratic nation.

The fate of Germany after the harsh peace terms of World War I and the subsequent rise of the Nazis should offer lesson enough on the potential ramifications of simple vengeance by the victor. Had the cry for retribution prevailed, we might today find ourselves with an implacable enemy instead of a friend, or a country so destabilized and adrift that it would have remained a social, political, and economic cripple for decades to come.

Summary

A terrible war was over and the nuclear age had begun.

Whether continued bombing and a blockade would have forced a Japanese surrender, making Hiroshima and Nagasaki unnecessary, remains a matter of conjecture. In the end, the threat of atomic horror, the Soviet entrance into the conflict, and the retention of the emperor were enough to get the Japanese to quit. And while some may have desired to try Hirohito for war crimes, the Allies decided otherwise, in hopes of bringing about a peaceful reconstruction.

America and Japan, bitter enemies for years, would now begin the healing process that would eventually produce a strong friendship of benefit to both nations.

Notes

Works cited and shortened titles can be found in full in the Bibliography.

Chapter 1. Seeds of Conflict

1. Time-Life Books, *WWII*, p. 158.
2. Ibid., p. 24.
3. Ibid., p. 158.
4. Ibid., p. 160.
5. Fuchida and Okumiya, *Midway*, p. 37.
6. Polmar, *Aircraft Carriers*, p. 45.
7. Willmott, *The Barrier and the Javelin*, p. 8.
8. Willmott, *Empires in the Balance*, p. 61.
9. Isom, *Midway Inquest*, pp. 32–33; Prange, *At Dawn We Slept*, pp. 167–169; Toland, *The Rising Sun*, p. 98.
10. Willmott, *The Barrier and the Javelin*, p. 5.
11. Willmott, *Empires in the Balance*, p. 66.

Chapter 2. Peace or War: 1941

1. Willmott, *Empires in the Balance*, p. 68.
2. Time-Life Books, *WWII*, pp. 160–161.
3. Polmar, *Aircraft Carriers*, p. 157.
4. Isom, *Midway Inquest*, p. 35.
5. Polmar, *Aircraft Carriers*, pp. 158–159.
6. Toland, *The Rising Sun*, pp. 208–209.
7. Isom, *Midway Inquest*, p. 53; Toland, *The Rising Sun*, p. 67; Willmott, *Empires in the Balance*, pp. 55–56.
8. Isom, *Midway Inquest*, p. 55; Toland, *The Rising Sun*, p. 75.
9. Parshall and Tully, *Shattered Sword*, p. 24; Willmott, Haruo, and Johnson, *Pearl Harbor*, p. 9.
10. Polmar, *Aircraft Carriers*, p. 146.

Chapter 3. December 7: Pearl Harbor

1. Prange, *At Dawn We Slept*, p. 184.
2. Ibid., pp. 299–300.
3. Polmar, *Aircraft Carriers*, p. 158; Time-Life Books, *WWII*, p. 161.
4. Time-Life Books, *WWII*, pp. 162–166.
5. Layton, *And I Was There*, p. 162; Prange, *At Dawn We Slept*, pp. 248–249; Willmott, Haruo, and Johnson, *Pearl Harbor*, p. 21.
6. Prange, *At Dawn We Slept*, p. 355.
7. Layton, *And I Was There*, p. 215; Prange, *At Dawn We Slept*, p. 406; Toland, *The Rising Sun*, pp. 199–200.
8. Prange, *At Dawn We Slept*, pp. 494–495, p. 567; Toland, *The Rising Sun*, p. 232.
9. Prange, *At Dawn We Slept*, pp. 495–501; Time-Life Books, *WWII*, p. 162; Toland, *The Rising Sun*, pp. 237–238; Willmott, Haruo, and Johnson, *Pearl Harbor*, p. 27.
10. Parshall and Tully, *Shattered Sword*, p. 13; Prange, *At Dawn We Slept*, p. 545.
11. Prange, *At Dawn We Slept*, pp. 542–544, 548.
12. Isom, *Midway Inquest*, pp. 66–67; Polmar, *Aircraft Carriers*, pp. 174–175; Willmott, Haruo, and Johnson, *Pearl Harbor*, p. 56.
13. Prange, *At Dawn We Slept*, p. 549.
14. Ewing, *USS* Enterprise, p. 14; Polmar, *Aircraft Carriers*, p. 164; Stafford, *The Big "E,"* p. 21; Tillman, *The Dauntless Dive Bomber of World War Two*, p. 20.
15. Isom, *Midway Inquest*, pp. 57–58.
16. Prange, *At Dawn We Slept*, p. 97.
17. Willmott, *The Barrier and the Javelin*, p. 169.
18. Stephan, *Hawaii under the Rising Sun*, pp. 117–118. Stephan specifies that Japanese plans formulated in September 1942 envisioned the use of the 2nd, 7th, and 53rd Divisions, along with an independent engineer regiment and a tank regiment.
19. Ibid., p. 99.
20. Polmar, *Aircraft Carriers*, p. 178; Prange, Goldstein, and Dillon, *Miracle at Midway*, pp. 10–11.

Chapter 4. Rising Sun: December 1941–April 1942

1. Polmar, *Aircraft Carriers*, p. 180.
2. Time-Life Books, *WWII*, p. 167.
3. Lundstrom, *The First Team*, p. 42.
4. Layton, *And I Was There*, pp. 345–346; Lundstrom, *Black Shoe Carrier Admiral*, pp. 39–41; Polmar, *Aircraft Carriers*, pp. 180–181; Prange, Goldstein, and Dillon, *Miracle at Midway*, pp. 5–6.
5. Burton, *Fortnight of Infamy*, p. 151; Isom, *Midway Inquest*, p. 72; Polmar, *Aircraft Carriers*, p. 176; Time-Life Books, *WWII*, p. 167; Toland, *The Rising Sun*, pp. 266–269.
6. Burton, *Fortnight of Infamy*, pp. 122–124.
7. Ibid., p. 129.

8. Isom, *Midway Inquest*, p. 72; Time-Life Books, *WWII*, pp. 166–170.
9. Willmott, *Empires in the Balance*, pp. 346–351.
10. Time-Life Books, *WWII*, pp. 170–172.
11. Willmott, *Empires in the Balance*, pp. 444–445.
12. Willmott, *The Barrier and the Javelin*, p. 33.
13. Douglas V. Smith, *Carrier Battles*, pp. 45–47.
14. Time-Life Books, *WWII*, p. 174.

Chapter 5. Turning the Tide: Spring 1942

1. Willmott, *Empires in the Balance*, p. 460.
2. Cohen, *Destination: Tokyo*, pp. 83–84; Glines, *The Doolittle Raid*, pp. 216–218; Layton, *And I Was There*, p. 388; Parshall and Tully, *Shattered Sword*, p. 43; Polmar, *Aircraft Carriers*, p. 209.
3. Ewing, *USS* Enterprise, p. 29; Parshall and Tully, *Shattered Sword*, p. 61; Stafford, *The Big "E,"* p. 79.
4. Peter C. Smith, *Midway: Dauntless Victory*, p. 50; Willmott, *The Barrier and the Javelin*, p. 210.
5. Layton, *And I Was There*, p. 390; Polmar, *Aircraft Carriers*, p. 212.
6. Lundstrom, *The First Team*, p. 182; Lundstrom, *Black Shoe Carrier Admiral*, p. 157; Douglas V. Smith, *Carrier Battles*, pp. 59–60; Willmott, *The Barrier and the Javelin*, p. 236.
7. Cressman, *That Gallant Ship*, p. 101.
8. Lundstrom, *The First Team*, p. 278.
9. Cressman, *That Gallant Ship*, p. 112; Henry, *Battle of the Coral Sea*, p. 84; Lundstrom, *The First Team*, pp. 284–285; Polmar, *Aircraft Carriers*, pp. 214–222; Time-Life Books, *WWII*, p. 177.
10. Parshall and Tully, *Shattered Sword*, p. 33.
11. Isom, *Midway Inquest*, p. 92; Lord, *Incredible Victory*, p. 5; Parshall and Tully, *Shattered Sword*, p. 37.
12. Parshall and Tully, *Shattered Sword*, p. 60.
13. Peter C. Smith, *Midway: Dauntless Victory*, p. 31.
14. Willmott, *The Barrier and the Javelin*, p. 304.
15. Layton, *And I Was There*, pp. 411–412; Lord, *Incredible Victory*, pp. 20–25; Parshall and Tully, *Shattered Sword*, pp. 92–93; Polmar, *Aircraft Carriers*, p. 226; Time-Life Books, *WWII*, pp. 252–253.

Chapter 6. Decisive Battle: Midway I

1. Polmar, *Aircraft Carriers*, p. 225; Time-Life Books, *WWII*, p. 253.
2. Time-Life Books, *WWII*, p. 253.
3. Lord, *Incredible Victory*, p. 11; Parshall and Tully, *Shattered Sword*, p. 63; Polmar, *Aircraft Carriers*, p. 226.
4. Willmott, *The Barrier and the Javelin*, p. 100.
5. Thomas B. Buell, *The Quiet Warrior*, p. 137; Prange, Goldstein, and Dillon, *Miracle at Midway*, pp. 99–100; Peter C. Smith, *Midway: Dauntless Victory*, p. 44.

6. Cressman, *That Gallant Ship*, p. 122; Lord, *Incredible Victory*, pp. 36–39; Polmar, *Aircraft Carriers*, p. 227; Prange, Goldstein, and Dillon, *Miracle at Midway*, pp. 100–101, 118–119.
7. Lundstrom, *Black Shoe Carrier Admiral*, p. 241.
8. Parshall and Tully, *Shattered Sword*, pp. 152–156.
9. Lundstrom, *Black Shoe Carrier Admiral*, pp. 254–255; Polmar, *Aircraft Carriers*, pp. 232–234.
10. Layton, *And I Was There*, p. 406; Lord, *Incredible Victory*, p. 40; Parshall and Tully, *Shattered Sword*, p. 99; Willmott, *The Barrier and the Javelin*, pp. 346–347.
11. Polmar, *Aircraft Carriers*, p. 231.
12. Isom, *Midway Inquest*, pp. 96–99; Layton, *And I Was There*, p. 435; Lord, *Incredible Victory*, pp. 42–44; Parshall and Tully, *Shattered Sword*, p. 99; Willmott, *The Barrier and the Javelin*, pp. 347–349.
13. Polmar, *Aircraft Carriers*, p. 232.
14. Isom, *Midway Inquest*, pp. 111–112; Polmar, *Aircraft Carriers*, p. 234.
15. Fuchida and Okumiya, *Midway*, p. 274.

Chapter 7. Rising Sun Eclipsed: Midway II

1. Polmar, *Aircraft Carriers*, p. 236.
2. Polmar, *Aircraft Carriers*, pp. 237–238; Time-Life Books, *WWII*, p. 256.
3. Gay, *Sole Survivor*, p. 119.
4. Polmar, *Aircraft Carriers*, p. 238.
5. Lord, *Incredible Victory*, pp. 85–86; Mrazek, *A Dawn Like Thunder*, p. 111; Peter C. Smith, *Midway: Dauntless Victory*, p. 95.
6. Mrazek, *A Dawn Like Thunder*, p. 166.
7. Ibid., p. 181.
8. Douglas V. Smith, *Carrier Battles*, p. 128.
9. Polmar, *Aircraft Carriers*, p. 237; Tillman, *The Dauntless Dive Bomber of World War Two*, pp. 68–69.
10. Thomas B. Buell, *The Quiet Warrior*, pp. 164–165; Mrazek, *A Dawn Like Thunder*, p. 166; Polmar, *Aircraft Carriers*, p. 250.
11. Cressman et al., *"A Glorious Page in Our History,"* p. 96; Lord, *Incredible Victory*, p. 153; Parshall and Tully, *Shattered Sword*, p. 217; Polmar, *Aircraft Carriers*, p. 239; Prange, Goldstein, and Dillon, *Miracle at Midway*, p. 260; Peter C. Smith, *Midway: Dauntless Victory*, p. 136; Tillman, *The Dauntless Dive Bomber of World War Two*, p. 70.
12. Cressman, et al., *"A Glorious Page in Our History,"* pp. 101–104; Lord, *Incredible Victory*, pp. 165–167; Parshall and Tully, *Shattered Sword*, p. 239; Peter C. Smith, *Midway: Dauntless Victory*, p. 139.
13. Lundstrom, *Black Shoe Carrier Admiral*, p. 261; Parshall and Tully, *Shattered Sword*, pp. 234–242; Polmar, *Aircraft Carriers*, pp. 238, 243; Prange, Goldstein, and Dillon, *Miracle at Midway*, pp. 262–264, 269–270; Stafford, *The Big "E,"* p. 95.

14. Cressman et al., "*A Glorious Page in Our History,*" p. 89; Cressman, *That Gallant Ship*, p. 125; Lord, *Incredible Victory*, pp. 154–155; Lundstrom, *Black Shoe Carrier Admiral*, pp. 250–251; Peter C. Smith, *Midway: Dauntless Victory*, p. 120; Tillman, *The Dauntless Dive Bomber of World War Two*, p. 71.
15. Parshall and Tully, *Shattered Sword*, pp. 324–327.
16. Thomas B. Buell, *The Quiet Warrior*, pp. 154–155; Lord, *Incredible Victory*, p. 256; Stafford, *The Big "E,"* pp. 101–102; Willmott, *The Barrier and the Javelin*, pp. 470–471.
17. Time-Life Books, *WWII*, p. 259.
18. Parshall and Tully, *Shattered Sword*, p. 48; Prange, Goldstein, and Dillon, *Miracle at Midway*, pp. 73–77.
19. Time-Life Books, *WWII*, p. 259.

Chapter 8. Island of Death: Guadalcanal

1. Time-Life Books, *WWII*, p. 259; Toland, *The Rising Sun*, pp. 396–397.
2. Time-Life Books, *WWII*, p. 259.
3. Ibid., pp. 260–261.
4. Lundstrom, *Black Shoe Carrier Admiral*, pp. 381–383; Polmar, *Aircraft Carriers*, p. 284; Time-Life Books, *WWII*, p. 260; Toland, *The Rising Sun*, p. 407.
5. Isom, *Midway Inquest*, pp. 244–245; Stafford, *The Big "E,"* p. 127; Toland, *The Rising Sun*, p. 413.
6. Toland, *The Rising Sun*, pp. 452–453.
7. Harold L. Buell, *Dauntless Helldivers*, p. 151; Polmar, *Aircraft Carriers*, pp. 291–292.
8. Time-Life Books, *WWII*, pp. 260–261.
9. Isom, *Midway Inquest*, pp. 244–245; Lundstrom, *Black Shoe Carrier Admiral*, p. 391; Polmar, *Aircraft Carriers*, p. 283; Douglas V. Smith, *Carrier Battles*, pp. 159–160; Time-Life Books, *WWII*, p. 260.
10. Harold L. Buell, *Dauntless Helldivers*, p. 161; Polmar, *Aircraft Carriers*, pp. 303–304; Tillman, *The Dauntless Dive Bomber of World War Two*, p. 111.
11. Polmar, *Aircraft Carriers*, pp. 286–289.
12. Ibid., p. 290.
13. Harold L. Buell, *Dauntless Helldivers*, p. 150; Ewing, *USS* Enterprise, pp. 48–49; Isom, *Midway Inquest*, p. 261; Polmar, *Aircraft Carriers*, pp. 292–299.

Chapter 9. Beginning of the End: 1943–1944

1. Time-Life Books, *WWII*, pp. 159–160.
2. Polmar, *Aircraft Carriers*, pp. 307–308; Time-Life Books, *WWII*, p. 263; Toland, *The Rising Sun*, pp. 500–503.
3. Koga was killed March 31, 1944, when his plane crashed. He was replaced by Admiral Soemu Toyoda.
4. Polmar, *Aircraft Carriers*, p. 359; Time-Life Books, *WWII*, p. 267.
5. Time-Life Books, *WWII*, p. 267.
6. Ibid., p. 268.

7. Thomas B. Buell, *The Quiet Warrior*, p. 227; Time-Life Books, *WWII*, pp. 269–270; Toland, *The Rising Sun*, pp. 533–534.
8. Harold L. Buell, *Dauntless Helldivers*, p. 251; Douglas V. Smith, *Carrier Battles*, pp. 235–236.
9. Tillman, *Clash of the Carriers*, p. 286.
10. Thomas B. Buell, *The Quiet Warrior*, pp. 251–252; Polmar, *Aircraft Carriers*, pp. 395–398; Tillman, *Clash of the Carriers*, p. 261; Time-Life Books, *WWII*, p. 270.
11. Polmar, *Aircraft Carriers*, p. 402; Tillman, *Clash of the Carriers*, p. 285.
12. Thomas B. Buell, *The Quiet Warrior*, pp. 286, 300, 301.
13. Time-Life Books, *WWII*, p. 273; Toland, *The Rising Sun*, pp. 573–574.
14. Time-Life Books, *WWII*, pp. 383–384.
15. Ibid.
16. Polmar, *Aircraft Carriers*, p. 426; Time-Life Books, *WWII*, p. 384.
17. Polmar, *Aircraft Carriers*, p. 427; Toland, *The Rising Sun*, pp. 634–635; Tully, *Battle of Surigao Strait*, p. 273.
18. Polmar, *Aircraft Carriers*, pp. 433–434; Time-Life Books, *WWII*, p. 386; Toland, *The Rising Sun*, pp. 639–640.
19. Polmar, *Aircraft Carriers*, p. 435; Time-Life Books, *WWII*, pp. 386–387; Toland, *The Rising Sun*, pp. 644–645.
20. Polmar, *Aircraft Carriers*, p. 432; Time-Life Books, *WWII*, p. 387; Toland, *The Rising Sun*, pp. 645–647.

Chapter 10. A Terrible Climax: 1945

1. Polmar, *Aircraft Carriers*, p. 463; Time-Life Books, *WWII*, pp. 396–397.
2. Time-Life Books, *WWII*, pp. 397–399.
3. Time-Life Books, *WWII*, pp. 399–401; Toland, *The Rising Sun*, p. 756.
4. Polmar, *Aircraft Carriers*, p. 469; Time-Life Books, *WWII*, 396.
5. Time-Life Books, *WWII*, pp. 401–403.
6. Time-Life Books, *WWII*, pp. 404–405; Toland, *The Rising Sun*, p. 820.
7. Time-Life Books, *WWII*, pp. 424–425.
8. Ibid., pp. 426, 430.
9. Leahy, *I Was There*, pp. 513–514.
10. Eisenhower, *The White House Years*, pp. 312–313.
11. Toland, *The Rising Sun*, p. 988.
12. Ibid., p. 989.

Bibliography

Alexander, Joseph H. *Utmost Savagery: The Three Days of Tarawa*. Annapolis, MD: Naval Institute Press, 1995.

Buell, Harold L. *Dauntless Helldivers*. New York: Orion Books, 1991.

Buell, Thomas B. *The Quiet Warrior: A Biography of Admiral Raymond A. Spruance*. Annapolis, MD: Naval Institute Press, 1987.

Burton, John. *Fortnight of Infamy: The Collapse* of *Allied Airpower West of Pearl Harbor*. Annapolis, MD: Naval Institute Press, 2006.

Cohen, Stan. *Destination: Tokyo*. Missoula, MT: Pictorial Histories Publishing Co., 1983.

Cressman, Robert J. *That Gallant Ship: USS* Yorktown. Missoula, MT: Pictorial Histories Publishing Co., 1985.

Cressman, Robert J., Steve Ewing, Barrett Tillman, Mark Horan, Clark Reynolds, and Stan Cohen. *"A Glorious Page in Our History": The Battle of Midway, 4–6 June 1942*. Missoula, MT: Pictorial Histories Publishing Co., 1990.

Eisenhower, Dwight D. *The White House Years: Mandate for Change, 1953–1956*. London: Heinemann, 1963.

Ewing, Steve. *USS* Enterprise *(CV-6): The Most Decorated Ship of World War II: An Illustrated History*. Missoula, MT: Pictorial Histories Publishing Co., 1982.

Fuchida, Mitsuo, and Masatake Okumiya. *Midway: The Battle that Doomed Japan: The Japanese Navy's Story*. Annapolis, MD: Naval Institute Press, 1955.

Gay, George, Jr. *Sole Survivor*. Naples, FL: privately printed, 1980.

Gingrich, Newt, and William R. Forstchen. *Pearl Harbor*. New York: St. Martin's Press, 2007.

Glines, Carroll V. *The Doolittle Raid: America's Daring First Strike against Japan*. New York: Orion Books, 1988.

Heinrichs, Waldo. *Threshold of War: Franklin D. Roosevelt and American Entry into World War II*. New York: Oxford University Press, 1990.

Henry, Chris. *Battle of the Coral Sea*. Annapolis, MD: Naval Institute Press, 2003.

Isom, Dallas W. *Midway Inquest: Why the Japanese Lost the Battle of Midway*. Bloomington, IN: Indiana University Press, 2007.

Layton, Edwin T. *And I Was There: Pearl Harbor and Midway: Breaking the Secrets*. New York: Quill/William Morrow, 1985.

Leahy, William D. *I Was There: The Personal Story of the Chief of Staff to Presidents Roosevelt and Truman*. London: Victor Gollancz, 1950.

Lord, Walter. *Incredible Victory*. New York: Harper & Row Publishers, 1967.

Lundstrom, John B. *Black Shoe Carrier Admiral*. Annapolis, MD: Naval Institute Press, 2006.

———. *The First Team: Pacific Naval Air Combat from Pearl Harbor to Midway*. Annapolis, MD: Naval Institute Press, 1984.

Miller, Edward S. *Bankrupting the Enemy: The U.S. Financial Siege of Japan before Pearl Harbor*. Annapolis, MD: Naval Institute Press, 2007.

Mrazek, Robert J. *A Dawn Like Thunder: The True Story of Torpedo Squadron Eight*. New York: Little, Brown and Company, 2008.

Parshall, Jonathan, and Anthony Tully. *Shattered Sword: The Untold Story of the Battle of Midway*. Washington, DC: Potomac Books, 2005.

Polmar, Norman. *Aircraft Carriers: A History of Carrier Aviation and Its Influence on World Events*, Vol. I, *1909–1945*. Washington, DC: Potomac Books, 2006.

Prange, Gordon W. *At Dawn We Slept: The Untold Story of Pearl Harbor*. New York: McGraw-Hill, 1981.

Prange, Gordon W., with Donald M. Goldstein and Katherine V. Dillon. *Miracle at Midway*. New York: McGraw-Hill, 1982.

Smith, Douglas V. *Carrier Battles: Command Decision in Harm's Way*. Annapolis, MD: Naval Institute Press, 2006.

Smith, Peter C. *Midway: Dauntless Victory*. Barnsley, South Yorkshire, UK: Pen & Sword Books Limited, 2007.

Stafford, Edward P. *The Big "E": The Story of the USS* Enterprise. New York: Random House, 1962.

Stephan, John. *Hawaii under the Rising Sun*. Honolulu, HI: University of Hawaii Press, 1984.

Tillman, Barrett. *Clash of the Carriers: The True Story of the Marianas Turkey Shoot of World War II*. New York: New American Library, 2005.

———. *The Dauntless Dive Bomber of World War Two*. Annapolis, MD: Naval Institute Press, 1976.

Time-Life Books. *WWII: Time-Life Books History of the Second World War*. New York: Prentice Hall Press, 1989.

Toland, John. *The Rising Sun: The Decline and Fall of the Japanese Empire*. New York: Random House, 1970.

Tully, Anthony P. *Battle of Surigao Strait*. Bloomington, IN: Indiana University Press, 2009.

Willmott, H. P. *The Barrier and the Javelin: Japanese and Allied Pacific Strategies, February to June 1942*. Annapolis, MD: Naval Institute Press, 1983.

———. *Empires in the Balance: Japanese and Allied Pacific Strategies to April 1942*. Annapolis, MD: Naval Institute Press, 1982.

Willmott, H. P., with Tohmatsu Haruo and W. Spencer Johnson. *Pearl Harbor*. London: Cassell, 2001.

About the Contributors

DONALD K. ALLEN is the author of *Tarawa: The Aftermath* (2001), the story of Tarawa Atoll before and after the November 20–23, 1943, assault by U.S. Marines. A book-related site continues with more veterans' stories and information, http://www.tarawatheaftermath.com.

KEITH ALLEN is a civilian analyst for the Department of the Navy and an amateur student of naval history and World War II.

MICHAEL BARNHART is Distinguished Teaching Professor of history at the State University of New York at Stony Brook. His published works include *Japan Prepares for Total War* (1988) and *Japan and the World since 1868* (1995). He is currently working on a survey of American foreign relations, from Jamestown to 9/11.

WILLIAM H. BARTSCH has been combining human resources planning consultancy work in developing countries with research and writing on the early years of the Pacific War since his retirement from the UN system in 1992. He is the author of *Doomed at the Start* (1992), *December 8, 1941: MacArthur's Pearl Harbor* (2003), and *Every Day a Nightmare* (2010).

JOHN BURTON is a former naval systems engineer who has spent much of his life studying the aviation industry and aviation history. During his twenty-year career with IBM as a business process analyst and sales and marketing executive, John consulted with many of the world's leading airframe manufacturers.

He holds degrees in manufacturing management and industrial engineering from California Polytechnic University and now works as an independent management consultant, living and writing in Irvine, California. *Fortnight of Infamy: The Collapse of Allied Airpower West of Pearl Harbor*, published by the Naval Institute Press (2006), is Mr. Burton's first book.

C. PETER CHEN received his MBA degree from Rutgers University. He is the founder and managing editor of the Web site World War II Database, http://ww2db.com, and is a staff member of the Web site Imperial Japanese Navy page, http://www.combinedfleet.com.

CLAYTON K. S. CHUN, PhD, is a faculty member and department chair at the U.S. Army War College in Carlisle Barracks, Pennsylvania. He is a retired Air Force officer and has written on a number of subjects related to national security, military history, and economics.

HAL FRIEDMAN is the associate chair of the Department of History and professor of modern history at Henry Ford Community College in Dearborn, Michigan. He additionally teaches part-time online for the U.S. Marine Corps University and Norwich University, as well as guest-lecturing for the U.S. Naval War College. He has published a trilogy on U.S. national security policy in the Pacific Basin in the 1940s, and is currently working on a trilogy on the U.S. Naval War College's reaction to the lessons of the Pacific War.

JOHN V. GARDNER joined the Marine Corps at the age of eighteen in 1940 and, after boot camp, was assigned to the 6th Defense Battalion. He specialized in maintaining telephones for the battalion's artillery. Sent to Midway in the summer of 1941, Gardner and his battalion skirmished with two Japanese destroyers on December 7, and defended Midway during the Japanese attack against the island in June 1942.

HAROLD J. GOLDBERG holds the David E. Underdown Chair in History at the University of the South in Sewanee, Tennessee. In addition to his course on World War II, Dr. Goldberg currently serves as chair of the Asian Studies Program. He is the author of two books on World War II, *Competing Voices from World War II in Europe: Fighting Words* (2010) and *D-day in the Pacific: The Battle of Saipan* (2007), as well as four volumes in a documentary study entitled *Documents of Soviet-American Relations* (1993–2001).

DONALD M. GOLDSTEIN teaches public and international affairs at the University of Pittsburgh. With Katherine V. Dillon and the late Gordon W. Prange, he created numerous World War II classics, including *At Dawn We Slept* (1982), *Miracle at Midway* (1983), and *God's Samurai: Lead Pilot at Pearl Harbor* (2003). Goldstein's research areas include history, public administration, political science, arms control, national interest and national security, theory and practice of international affairs, policy analysis, foreign policy process, international relations, and administrative theory. He is one of the nation's foremost experts in military history, particularly World War II and the Korean conflict. He is regularly cited in national publications and has appeared in television documentaries about military history and warfare.

JAMES H. HALLAS is a military historian who has authored books about the U.S. Marine Corps in World War II and the American Expeditionary Forces in World War I. He lives in Connecticut.

LEWIS A. HOPKINS entered the U.S. Navy Reserve as an apprentice seaman in July 1940 and retired as a rear admiral in July 1974. He was designated a naval aviator in September 1941 and saw action in the Pacific theater of World War II, participating in the Battle of Midway in June 1942. After the war, he served in many aviation-related activities, including research and development, and served in many political positions, including mayor of Del Mar, California.

TIMOTHY H. JACKSON is a retired naval officer and professor of strategy and policy at the U.S. Naval War College in Newport, Rhode Island. He is the director of the College of Distance Education and director of Academic Support, serving more than 40,000 students from all branches of the military services as well as students from other federal agencies.

ALVIN KERNAN was born in Manchester, Georgia, in 1923. He grew up on a ranch in Wyoming, enlisted in the Navy in 1941, and was on board the USS *Enterprise* at Pearl Harbor. An ordnanceman and aerial gunner in torpedo planes, he served on board aircraft carriers in the Pacific for the remainder of the war. After the war, he taught at Yale and Princeton for many years. The best-known of his writings is his highly praised *Crossing the Line*, a memoir of his Navy life, published in 1995. Yale University Press brought out a new edition in 2006.

NORMAN J. "DUSTY" KLEISS graduated from the U.S. Naval Academy in 1938, became qualified as a carrier pilot after graduation from Pensacola and Oppa Locka, and joined Scouting 6 of the USS *Enterprise* in May 1941. He was in combat as an SBD dive-bomber pilot from the Japanese attack on Pearl Harbor through the Battle of Midway. He bombed the *Kaga*, *Hiryu*, and *Mikuma*, and received a Navy Cross, in addition to the Distinguished Flying Cross he previously received for his bombing of a cruiser in the Marshall Islands. He taught hundreds of young aviators how to dive-bomb. As the officer in charge of the Catapult and Arresting Gear Division of the Bureau of Aeronautics, he gained approval for replacing hydraulic catapults with steam catapults, and replacing the landing signal officer, at the stern of a carrier, with the mirror landing system.

YOJI KODA is a graduate of Japan National Defense Academy (1972), the Japan Maritime Self-Defense Force (JMSDF) Officer Candidate School at Etajima, and JMSDF Staff College, and, in 1992, the U.S. Naval War College. As a vice admiral he commanded the Fleet Escort Force (2003–2004), later serving as director general of the Joint Staff Office (2004–2005), commandant of the Sasebo JMSDF District, and as commander in chief, Self-Defense Fleet, from 2007 until his retirement in 2008. He has written widely on history and security in both Japanese and English, including "The Russo-Japanese War: Primary Causes of Japanese Success" (*Naval War College Review*, Spring 2003) and "The Emerging Republic of Korea Navy: A Japanese Perspective" (*Naval War College Review*, Spring 2010). Vice Admiral Koda is currently serving as a senior fellow at Asia Center, Harvard University, until summer 2011.

JOHN LUNDSTROM is a curator emeritus of history at the Milwaukee Public Museum, where he worked for thirty-seven years. He is the author of several books on the Pacific War, including *Black Shoe Carrier Admiral* (2006), *The First Team* (1984), and *The First Team and the Guadalcanal Campaign* (1994).

LEX McAULAY served for twenty-two years in the Australian Regular Army, with more time in the Reserves, and had three tours of duty in South Vietnam, which allowed him to experience the war from infantry squad level to General William Westmoreland's HQ. He is the author of a number of books on Australian and U.S. military events in World War II and Vietnam, published in Australia, the United Kingdom, and the United States, including *MacArthur's Eagles* (2005). He lives in eastern Australia.

ROBERT MRAZEK was a five-term Democratic member of the United States House of Representatives, representing New York's Third Congressional District on Long Island for most of the 1980s. He is the author of five books, including *A Dawn Like Thunder: The True Story of Torpedo Squadron 8 in the Pacific War* (2008), *Stonewall's Gold* (2000), and *Deadly Embrace* (2006).

JON PARSHALL is co-author of *Shattered Sword: The Untold Story of the Battle of Midway* (2007), and is the longtime maintainer of the Web site Imperial Japanese Navy Page (http://www.combinedfleet.com). He has published numerous articles on the topic of the Imperial Navy, appeared on both the Discovery Channel and the History Channel, and has spoken at venues such as the U.S. Naval War College, the 65th Pearl Harbor Memorial Symposium, the Nimitz Museum, and the Pritzker Military History Library. He lives in Minneapolis with his wife and two children

STEPHEN D. REGAN received a bachelor's degree in history and English from Upper Iowa University, and subsequently earned master's and doctorate degrees from Winona State University (Minnesota) and the University of South Dakota respectively. After a stint in the Navy and several years in education, he was invited back to Upper Iowa University as professor and academic dean. Among his articles and books are *In Bitter Tempest: The Biography of Admiral Frank Jack Fletcher* (1994) and *Pioneering Spirit: The History of Upper Iowa University* (2008). He is retired and living in Cedar Rapids, Iowa.

GORDON ROTTMAN has served for twenty-six years in the U.S. Army in Special Forces, airborne infantry, long-range reconnaissance patrol, and military intelligence assignments in the regular Army, Army National Guard, and Army Reserve. He was a special operations forces scenario writer for the Army's Joint Readiness Training Center for thirteen years. Gordon began writing military history books in 1984 and has since published over one hundred books. He currently lives in Cypress, Texas.

RONALD W. RUSSELL, a retired chief warrant officer in the Naval Reserve, serves as editor for the Battle of Midway Roundtable, an international association of veterans, historians, authors, and others having a strong focus on that battle. He is the author of *No Right to Win: A Continuing Dialogue with Veterans of the Battle of Midway (2006)*, and his articles have appeared in radio technology and naval history magazines.

WILLIAM SAGER was a second lieutenant in the U.S. Marine Corps in 1942, a rifle platoon leader in K Company of the 3d Battalion, 1st Marines. On August 7, 1942, the 3d Battalion, 1st Marines, under the command of Colonel William N. McKelvy Jr., landed on Guadalcanal, and Sager and his fellow Marines proceeded to fulfill the mission of their combat team.

NICHOLAS EVAN SARANTAKES is an associate professor of strategy and policy at the U.S. Naval War College. He is the author of two books on World War II, *Allies against the Rising Sun: The United States, the British Nations, and the Defeat of Imperial Japan* (2009) and *Seven Stars: The Okinawa Battle Diaries of Simon Bolivar Buckner Jr. and Joseph Stilwell* (2004). He also wrote a book on the aftermath of war, *Keystone: The American Occupation of Okinawa and U.S.-Japanese Relations* (2000).

FRANK SHIRER is the chief archivist for the U.S. Army Center of Military History at Fort McNair, Washington, D.C., specializing in World War II and the Soviet military. An army veteran with twenty years' enlisted and commissioned service in the U.S. Army, he has served in the field artillery and military intelligence, from battery level to the DCSINT (deputy chief of staff for intelligence) staff at the Pentagon. Since retiring from the Army, he has worked for the U.S. Army in a variety of positions, moving from historian to his current position as chief of the Center's archive.

DOUGLAS V. SMITH is professor of strategy and head of the Strategy and Policy Division at the U.S. Naval War College's College of Distance Education. He is a graduate of the U.S. Naval Academy, Naval Postgraduate School, and Naval War College, and holds a PhD in military history from Florida State University. He is also the author of the book *Carrier Battles* (2006).

PETER C. SMITH was born in Norfolk, England, in October 1940. He is a member of the Society of Authors, London, and the London Press Club. He has had sixty-eight books published worldwide in many different languages, mostly on factual subjects relating to World War II, including the recently much-lauded *Midway: Dauntless Victory* (2007), *Fist from the Sky* (2006), and *Pedestal: The Convoy that Saved Malta* (1999). He currently resides near Bedford, England.

FRANK SNYDER is a retired Navy captain, formerly on the faculty at the U.S. Naval War College, Newport, Rhode Island. As a professor emeritus there, he gives lectures about naval battles, particularly those of World War II in the Pacific.

JAY A. STOUT is a retired Marine Corps fighter pilot who flew thirty-seven combat missions during Operation Desert Storm. An award-winning military writer and historian, he is the author of many books and articles. He additionally works as a senior analyst in the defense industry.

ROBERT SWAN was an ensign serving as navigator on one of Midway's PBYs, the Catalina flying boats employed in the search for the approaching Japanese fleet in June 1942. Swan's PBY spotted the Japanese transport group on June 3. Swan also took part in scouting missions in the Solomon Island chain in 1943.

BARRETT TILLMAN is an award-winning author and historian with more than forty histories, novels, and biographies to his credit. A lifelong student of military aviation, he grew up flying historic aircraft with his father, a Navy-trained pilot of World War II. Tillman is best known for his operational histories of naval aircraft including the Dauntless, Hellcat, and Corsair. *On Yankee Station* (1987), his co-authored study of air operations in Vietnam, was chosen for the Marine Corps reading list and was used as a "reality check" by air wings embarked for Operation Desert Storm. He has appeared on CNN, National Geographic Channel, and the History Channel.

ANTHONY P. TULLY is an independent scholar and historian of the Pacific War and other historical and political subjects. He is co-author (with Jon Parshall) of *Shattered Sword* (2007), an acclaimed study of the Battle of Midway, and is the author of *Battle of Surigao Strait* (2009). He is the managing editor of the Web site Imperial Japanese Navy page, http://www.combinedfleet.com, and he lives in Texas.

SUMNER WHITTEN graduated from Amherst College in 1940 and enlisted in the Marines. Commissioned a second lieutenant in 1941, Whitten fought in the Battle of Midway in 1942 and in the campaign for Guadalcanal in 1942–43. He also served in the Korean War in 1951–52.

H. P. WILLMOTT holds a master's degree from Liverpool University and a doctorate from London University in the UK, and is a qualified staff officer (U.S. National War College, 1992–94). A fellow of the Royal Historical Society, he has served on the faculties of Temple University, the University of Memphis, and the Department of Military Strategy and Operations, National War College. He has written extensively on modern naval and military subjects, including *Empires in the Balance* (1982), *The Barrier and the Javelin* (1985), *When Men Lost Faith in Reason: Reflections on Warfare in the Twentieth Century* (2002), and *The Battle of Leyte Gulf: The Last Fleet Action* (2005), the last of which won the Society for Military History's top award for 2006.

VOLUME EDITOR: JIM BRESNAHAN is a broadcast journalist in Lexington, Virginia. He is the author of *Revisioning the Civil War* (2005), a book on how the history of the American Civil War might have been different. He also wrote *Play it Again: Baseball Experts on What Might Have Been* (2006), a book that focuses on how the history of baseball might have been altered. McFarland of Jefferson, North Carolina, published both books.

Index

The **Naval Institute Press** is the book-publishing arm of the U.S. Naval Institute, a private, nonprofit, membership society for sea service professionals and others who share an interest in naval and maritime affairs. Established in 1873 at the U.S. Naval Academy in Annapolis, Maryland, where its offices remain today, the Naval Institute has members worldwide.

Members of the Naval Institute support the education programs of the society and receive the influential monthly magazine *Proceedings* or the colorful bimonthly magazine *Naval History* and discounts on fine nautical prints and on ship and aircraft photos. They also have access to the transcripts of the Institute's Oral History Program and get discounted admission to any of the Institute-sponsored seminars offered around the country.

The Naval Institute's book-publishing program, begun in 1898 with basic guides to naval practices, has broadened its scope to include books of more general interest. Now the Naval Institute Press publishes about seventy titles each year, ranging from how-to books on boating and navigation to battle histories, biographies, ship and aircraft guides, and novels. Institute members receive significant discounts on the more than eight hundred Press books in print.

Full-time students are eligible for special half-price membership rates. Life memberships are also available.

For more information about Naval Institute Press books that are currently available, visit www.usni.org/press/books. To learn about joining the U.S. Naval Institute, please write to:

Member Services
U.S. NAVAL INSTITUTE
291 Wood Road
Annapolis, MD 21402-5034
Telephone: (800) 233-8764
Fax: (410) 571-1703
Web address: www.usni.org